CHEMISTRY RESEARCH AND APPLICATIONS

A COMPREHENSIVE GUIDE TO FORMALDEHYDE

CHEMISTRY RESEARCH AND APPLICATIONS

Additional books and e-books in this series can be found on Nova's website under the Series tab.

CHEMISTRY RESEARCH AND APPLICATIONS

A COMPREHENSIVE GUIDE TO FORMALDEHYDE

NATASJA A. BACH
EDITOR

Copyright © 2021 by Nova Science Publishers, Inc.

All rights reserved. No part of this book may be reproduced, stored in a retrieval system or transmitted in any form or by any means: electronic, electrostatic, magnetic, tape, mechanical photocopying, recording or otherwise without the written permission of the Publisher.

We have partnered with Copyright Clearance Center to make it easy for you to obtain permissions to reuse content from this publication. Simply navigate to this publication's page on Nova's website and locate the "Get Permission" button below the title description. This button is linked directly to the title's permission page on copyright.com. Alternatively, you can visit copyright.com and search by title, ISBN, or ISSN.

For further questions about using the service on copyright.com, please contact:
Copyright Clearance Center
Phone: +1-(978) 750-8400   Fax: +1-(978) 750-4470   E-mail: info@copyright.com.

NOTICE TO THE READER

The Publisher has taken reasonable care in the preparation of this book, but makes no expressed or implied warranty of any kind and assumes no responsibility for any errors or omissions. No liability is assumed for incidental or consequential damages in connection with or arising out of information contained in this book. The Publisher shall not be liable for any special, consequential, or exemplary damages resulting, in whole or in part, from the readers' use of, or reliance upon, this material. Any parts of this book based on government reports are so indicated and copyright is claimed for those parts to the extent applicable to compilations of such works.

Independent verification should be sought for any data, advice or recommendations contained in this book. In addition, no responsibility is assumed by the Publisher for any injury and/or damage to persons or property arising from any methods, products, instructions, ideas or otherwise contained in this publication.

This publication is designed to provide accurate and authoritative information with regard to the subject matter covered herein. It is sold with the clear understanding that the Publisher is not engaged in rendering legal or any other professional services. If legal or any other expert assistance is required, the services of a competent person should be sought. FROM A DECLARATION OF PARTICIPANTS JOINTLY ADOPTED BY A COMMITTEE OF THE AMERICAN BAR ASSOCIATION AND A COMMITTEE OF PUBLISHERS.

Additional color graphics may be available in the e-book version of this book.

Library of Congress Cataloging-in-Publication Data

ISBN: 978-1-53619-465-4

Published by Nova Science Publishers, Inc. † New York

CONTENTS

Preface   vii

Chapter 1   Thermal Curing Characteristics of Bisbenzoxazines Based on 4,4'-Diaminodiphenyl Sulfone and 4,4'-Dihydroxy Diphenyl Sulfone and Thermal Degradation of Thermosets Derived from these Monomers   1
S. Shamim Rishwana, S. Siva Kaylasa Sundari, Arunjunai Raj Mahendran and C. T. Vijayakumar

Chapter 2   Formaldehyde: Polymer Surface Chemistry and Detection   47
Sanjiv Sonkaria and Hyun-Joong Kim

Chapter 3   Electrochemical Assessment of Formaldehyde with Doped Nanorod Materials   87
Mohammed Muzibur Rahman, Abdullah M. Asiri, M. M. Alam and Jamal Uddin

Chapter 4	Thermal Studies on Styryl Modified Resorcinolic Resin - Formaldehyde Donor Adhesion Systems for Rubber Industry *J. Dhanalakshmi, S. Siva Kaylasa Sundari and C. T. Vijayakumar*	**109**
Index		**127**

PREFACE

This monograph comprises four chapters concerning formaldehyde, a colorless, flammable, strong-smelling chemical that is an essential component in many manufacturing industries. Chapter 1 explicates the synthesis of diaminodiphenyl sulphone bisbenzoxazine by three step process and the synthesis of bisphenol-S bisbenzoxazine by solvent method. Chapter 2 remarks on the substantial growth trajectory of formaldehyde, examining the potential health-related effects associated with its rise in popularity and discussing how nanotechnological advancement may play a future role in mitigating the chemical's harmful impact. Chapter 3 discusses the growth and development of doped nanostructure materials by using hydrothermal method in alkaline phase, concluding that the prepared doped nanomaterial could be implemented in a broad scale for efficient electro-chemical sensor applications for environmental and healthcare fields. Chapter 4 describes the production of reinforcing rubber products, specifically tires, using rubber compounded with methylene acceptor and methylene donor.

Chapter 1 - The diaminodiphenyl sulphone bisbenzoxazine (DDS-Bz) was synthesized by three step process and bisphenol-S bisbenzoxazine (BS-a) was synthesized by solvent method. The synthesized compounds were thermally polymerized and the structural characterization was done by using Fourier Transformation Infrared Spectral studies (FTIR). The thermal studies were carried out in Differential Scanning Calorimeter (DSC) and Thermogravimetric Analyzer (TGA). The apparent activation energies for the thermal curing of the monomers (Ea-C) and the degradation process (Ea-D) for the cured thermosets were calculated using different isoconversional kinetic methods. The variation in Ea-C noted for the thermal curing of these two bisbenzoxazines is attributed to the operation of different mechanism in the curing process and to the position of phenol and amine group present in the molecules. The variation of the Ea-D for the degradation of the polymers may be due to the structure and the volatile products formed during the degradation process. The cured thermosets are thermally more stable and leads to high char residue. The high average char residue was noted for the polymerized bisphenol-S based bisbenzoxazine (PBS-a). The volatile products obtained during the thermal degradation of the polymers were analyzed using thermogravimetric analyzer coupled to Fourier transform infrared spectrophotometer (TG-FTIR). Sulphur dioxide was found to be the major product and was released during the degradation. The thermal stability of the polymer is decided by the nature of the aromatic nucleus present in the bisbenzoxazine and the temperature at which the compound is polymerized.

Chapter 2 - The global market trend predicts that the growth trajectory of the chemical "formaldehyde" will substantially increase production marking an unprecedented global economic expansion. The chemical profile of formaldehyde particularly in view of its reactive and volatile nature raises much concern about the potential chemical effects on disease progression due to inhalation at multi-levels in both natural and biological environments at bulk and nanoscale dimensions. As the global dependency on the use formaldehyde surges with no signs of slowing, much effort has been directed to investigate health-related issues and how nanotechnological advancement might play a greater role in in reducing the harmful impact of the chemical. The issues addressed in this chapter will converge upon a discussion and review of environmental and biological implications of the industrial growth of formaldehyde and a comprehensive evaluation of measures addressing the potentialities through the development of sensitive detection technologies.

Chapter 3 - In this approach, the growth and development of doped nanostructure materials (i.e., nanorods) by using hydrothermal method in alkaline phase were discussed. The structural, optical and chemical properties of nanorods (NRs) were characterized using various methods such as UV/vis., FTIR, Raman spectroscopy, powder XRD, and FESEM, XEDS, XPS, etc. Doped $ZnFe_2O_4$ is an attractive nanorod for potential application in chemical sensing by easy and reliable electrochemical method, where formaldehyde is considered as a model compound. The chemical sensor performances are exhibited the higher sensitivity, good stability, and repeatability of the sensor enhanced significantly using doped NRs of thin-film with conducting coating binders on silver electrodes. The calibration plot is linear over the large dynamic range, where the sensitivity and detection limit were calculated based on signal/noise ratio ($\sim^{3N}/_S$) in short response time. At last, it is concluded that the prepared doped nanomaterial could be implemented in a broad-scale for an efficient electro-chemical sensor applications for environmental and healthcare fields.

Chapter 4 - Reinforcing rubber products were made using rubber compounded wih methylene acceptor and methylene donor. Industrial materials like resorcinol formaldehyde resins (RF) and hexamethoxymethylmelamine (HMMM) are used as methylene acceptor and methylene donor respectively. RF resin possesses excellent usable properties, but they have high hygroscopicity. This propery increases fuming at the time of rubber compounding and handling which is a inherent problem of RF resins to be used in rubbers. Styryl modified resorcinol formaldehyde resin (R2) and HMMM are used in tire industries as an enhanced rubber compounding material. The thermal studies of structurally modified resorcinol formaldehyde resins (alkyl/ aralkyl) are scarce. Styryl group reduces the hygroscopic property of RF resin and resin volatility due to the presence of lower levels of free resorcinol (1-5 %) in the R2. The curing characteristics of the pure R2, pure HMMM and R2:HMMM blends (80:20, 60:40, 50:50, 40:60 and 20:80 weight ratios) were investigated using Differential Scanning Calorimetry (DSC) and it was found that the increase in HMMM amount in the blends alter the curing behaviour of pure R2. The amount of heat liberated during the thermal curing of these blends decreased when HMMM content increased in the blends. Thermal properties of thermally cured materials were investigated using thermogravimetric analysis (TGA). TG results indicated that the polymers from the blends were thermally more stable compared to the pure resin R2. Among the materials investigated, the 60:40 and 50:50 weight ratio R2:HMMM blends showed the highest char yield (~36.0 %) at 750 °C. The Fourier Transform Infrared Spectroscopic (FTIR) studies of these materials revealed the formation of quinone methide structures during the reaction between the R2 and HMMM and the interaction existing between R2 and HMMM is explicit.

In: A Comprehensive Guide to Formaldehyde ISBN: 978-1-53619-465-4
Editor: Natasja A. Bach © 2021 Nova Science Publishers, Inc.

Chapter 1

THERMAL CURING CHARACTERISTICS OF BISBENZOXAZINES BASED ON 4,4'-DIAMINODIPHENYL SULFONE AND 4,4'-DIHYDROXY DIPHENYL SULFONE AND THERMAL DEGRADATION OF THERMOSETS DERIVED FROM THESE MONOMERS

S. Shamim Rishwana[1], PhD, S. Siva Kaylasa Sundari[1], Arunjunai Raj Mahendran[2], PhD and C. T. Vijayakumar[3,], PhD*

[1]Department of Chemistry, Kamaraj College of Engineering and Technology (Autonomous), S.P.G.C. Nagar, K. Vellakulam, India
[2]Kompetenzzentrum Holz GmbH (W3C), Klagenfurter Straße. St. Veit an der Glan, Austria

---

[*] Corresponding Author's E-mail: ctvijay22@yahoo.com.

[3]Department of Polymer Technology, Kamaraj College of Engineering and Technology (Autonomous), S.P.G.C. Nagar, K. Vellakulam, India

ABSTRACT

The diaminodiphenyl sulphone bisbenzoxazine (DDS-Bz) was synthesized by three step process and bisphenol-S bisbenzoxazine (BS-a) was synthesized by solvent method. The synthesized compounds were thermally polymerized and the structural characterization was done by using Fourier Transformation Infrared Spectral studies (FTIR). The thermal studies were carried out in Differential Scanning Calorimeter (DSC) and Thermogravimetric Analyzer (TGA). The apparent activation energies for the thermal curing of the monomers (Ea-C) and the degradation process (Ea-D) for the cured thermosets were calculated using different isoconversional kinetic methods. The variation in Ea-C noted for the thermal curing of these two bisbenzoxazines is attributed to the operation of different mechanism in the curing process and to the position of phenol and amine group present in the molecules. The variation of the Ea-D for the degradation of the polymers may be due to the structure and the volatile products formed during the degradation process. The cured thermosets are thermally more stable and leads to high char residue. The high average char residue was noted for the polymerized bisphenol-S based bisbenzoxazine (PBS-a). The volatile products obtained during the thermal degradation of the polymers were analyzed using thermogravimetric analyzer coupled to Fourier transform infrared spectrophotometer (TG-FTIR). Sulphur dioxide was found to be the major product and was released during the degradation. The thermal stability of the polymer is decided by the nature of the aromatic nucleus present in the bisbenzoxazine and the temperature at which the compound is polymerized.

Keywords: bisbenzoxazine, DSC, TGA, Ea-C, Ea-D, TG-FTIR

INTRODUCTION

Macromolecules or Polymers are very large molecules made up of a large number of small molecules, called monomers, covalently bonded together. This specific molecular structure (chain like structure) of polymeric materials is responsible for their intriguing mechanical properties. The process by which the monomer molecules are linked to form a big molecule is called polymerization (Flory 1953). Polymers are the material of choice in a vast range of applications such as packaging, building and construction, transportation, electrical and electronic equipment, agriculture, as well as in the medical and the sports sectors. Polymers include biopolymers, rubbers and plastics (thermoplastics and thermosets). Thermoset materials include polymers, resins and plastics which are infusible and insoluble and hence it exists in permanent solid state after curing. Polymers within the thermoset material cross-link during the curing process form a stable covalent bond. This means that thermosets will not melt even when exposed to high temperatures (Dodiuk H 2014). Polymers that can deal successfully in high temperature engineering environments are termed high performance polymers. High temperature polymers are leading the polymer market because of their variety of applications. They find widespread use in areas such as adhesives, structural applications in aerospace, printed circuit boards, conductive polymer elements and encapsulation materials for electronic applications (Meador 1998). The aerospace industry and space programs have created new demands for high temperature polymers which can withstand higher temperatures owing to their excellent thermal and thermo-oxidative stability, high char yield, good chemical inertness, abrasion resistance and flame retardancy (Heregenrother 2003). The best known members of the thermoset family are phenolic resins, amino resins, alkyd resins, unsaturated polyester resins, vinyl ester resins, allyl resins, epoxy resins, isocyanate derived polymers, bismaleimides (BMI) and cyanate esters. Hence, thermosets are widely used in everyday consumer products and engineering applications. They can provide design and manufacturing flexibility along with the combinations of properties.

Formaldehyde based thermoset resins plays an important role across the world. The resins include phenol-formaldehyde (PF) resins – phenoplasts and urea-formaldehyde (UF), melamine-formaldehyde (MF), melamine-urea formaldehyde (UF-MF) resins – aminoplasts. Traditional phenolic materials are the cross linked products of their low molecular weight precursors, either of novolac or resole type (Figure 1). These approaches have been successful in bringing about the first class of synthetic plastic materials since the turn of 20th century. The materials obtained exhibit good heat resistance, flame retardancy and dielectric properties. Thus, phenoplasts have been widely used in construction, household and electrical facilities. Another notable advantage is that the raw materials and the fabricating process are inexpensive. A book on phenolic materials has given thorough overview to this well-established materials (Knop et al. 1985).

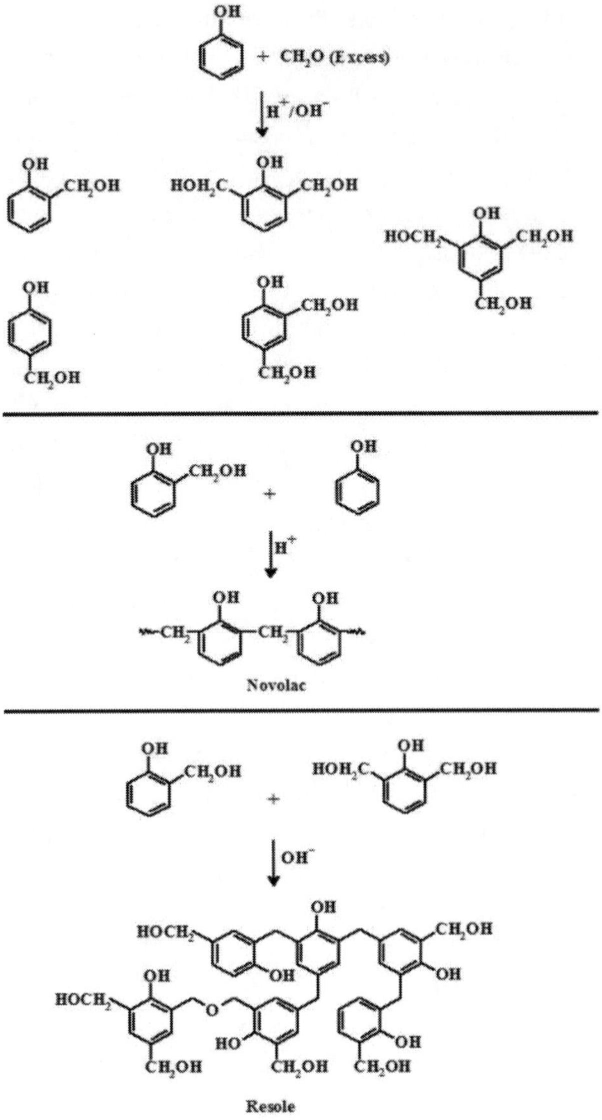

Figure 1. Synthesis of novolac and resole.

Amino resins or aminoplasts are prepared from the condensation of either urea with formaldehyde (UF resins) or melamine with formaldehyde (MF resins). UF resins are prepared by the nucleophilic addition of urea to formaldehyde to give methylol derivatives. Subsequent condensation of these derivatives gives the final high molecular weight resin (Figure 2).

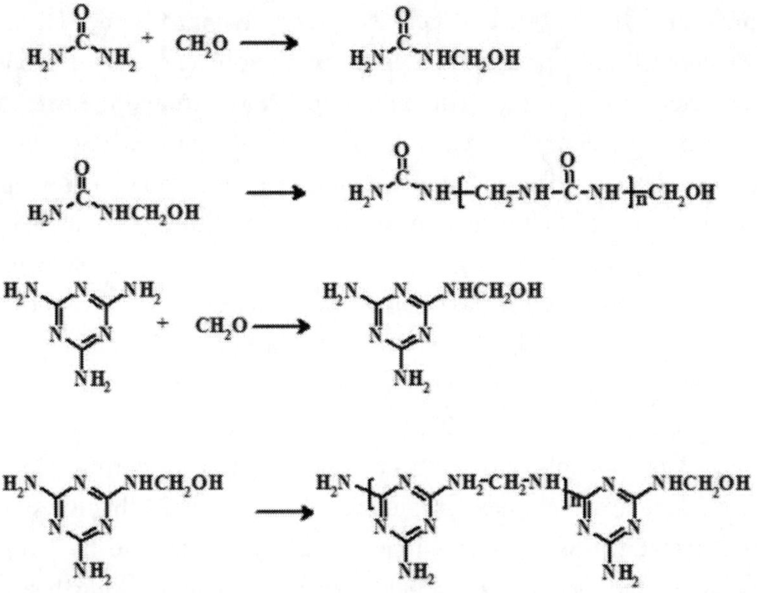

Figure 2. Synthesis and polymerization of amino resins.

Benzoxazine (BZ) are new class of high performance thermoset resins and it consists of single benzene ring fused to another six-membered heterocycle containing an oxygen atom and a nitrogen atom. There are a number of possible isomeric benzoxazines depending upon the relative position of the two hetero atoms and the degrees of oxidation of this oxazine ring system. The history of small molecular weight benzoxazines dates back to more than 60 years ago. Holly and Cope first reported the synthesis of benzoxazine in 1944 [Holly and Cope 1944]. From 1950's to 1960's, Burke and co-workers synthesized many benzoxazines (Bruke et al. 1949 and 1952) and naphthoxazines (Bruke et al. 1949 and 1952) for the purpose of antitumor activity test. Reiss and co-workers (Reiss et al.

1985) investigated the polymerization of mono-functional benzoxazines with and without phenol as an initiator, resulting in linear polymers under 4000 molecular weight. It was Schreiber (Schreiber 1973) in his patents in the early 1970s who reported the synthesis of small oligomers as a modifier for epoxy resin, though no details of the oligomer properties were reported. In the 1980s, Higginbottom in his quest to develop a coating system was the first to develop a cross linked polybenzoxazines based on a multifunctional benzoxazines, though again no polybenzoxazine properties were reported in his patents [Higginbottom 1985]. Around the same time, Riese et al. studied the reaction kinetics of benzoxazine oligomer formation using mostly the monofunctional benzoxazines, showing that large molecular weight linear polybenzoxazines cannot be obtained from monofunctional benzoxazines. Several monomers studied showed the molecular weights roughly between a few hundreds to a few thousands upon polymerization [Reiss 1985]. In 1988, Turpin and Thrane reported in their patent the self-curable benzoxazine functional cathodic electrocoat resin formulation [Turpin 1988]. They described the use of both multifunctional phenols and multifunctional amines as the raw materials for low molecular weight benzoxazine resins. It is in 1994 that the study on polybenzoxazine properties was reported for the first time by Ning and Ishida [Ning and Ishida 1996, 2000, 2001], despite its long history of discovery of low molecular weight benzoxazine in 1944, possibility of polymerizing it in 1973 and utilization of a cross linked polybenzoxazine in 1985.

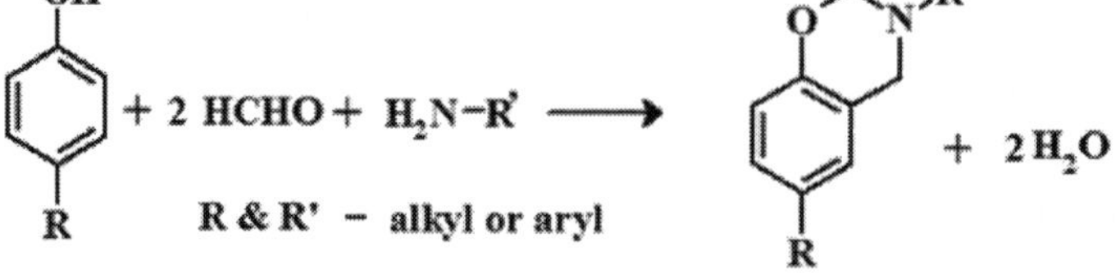

Figure 3. Synthesis of monobenzoxazine.

The precursors of polybenzoxazine are formed from phenols and formaldehyde in the presence of aliphatic or aromatic amines (Ishida et al.

1996). The choice of phenol and amine permits design flexibility and tailoring of polymer properties. Aromatic oxazines were synthesized through Mannich reactions from phenols, formaldehyde and amines (Figure 3).

Polybenzoxazines (PBZ) are synthesized by ring opening polymerization of the aromatic bisbenzoxazines. PBZ that can overcome many shortcomings associated with traditional phenolic resins have been synthesized and characterized extensively by Ishida and coworkers. They possess excellent resistance to chemicals (Kim and Ishida 2001), UV light (Macko and Ishida 2000) and high Tg values (Ishida and Rodriguez 1995). A wide variety of BZ monomers can be readily prepared by varying the precursor components by the Mannich condensation reaction. They are therefore of great interest in synthetic fields. Phenolic resins from such benzoxazines were shown to be cross linked polymers and exhibit good mechanical properties. Some benzoxazine resins were found to have near-zero shrinkage or volumetric expansion upon curing. Polybenzoxazines provide tremendous freedom in molecular design which can modify products based on the structure property relationships. Furthermore, they do not release byproducts during curing and no catalyst is needed.

The most investigated polybenzoxazine is derived from bisphenol-A and aniline. This benzoxazine is treated as a 'benchmark' among polybenzoxazine and the properties of new polybenzoxazine are compared with it. The material property balance of polybenzoxazines is excellent. Good thermal, chemical, electrical, mechanical and physical properties make them attractive alternatives to existing applications. Additional new applications can be developed by utilizing unique properties of polybenzoxazines that have not been often observed by other well-known polymers. Those properties include near-zero shrinkage, very high char yield, fast development of mechanical properties as a function of conversion, glass transitions much higher than curing temperatures, excellent electrical properties and low water uptake despite having many hydrophilic groups. They can be synthesized from inexpensive raw materials and polymerized by a ring-opening addition reaction, yielding no reaction by-product. The as-synthesized mixture consists of monomer and oligomers that contain phenolic groups. For practical applications, the mixture is sufficient but for controlled structure and properties, the monomer is freed of the oligomers. Shamim Rishwana and her coworkers studied the curing and thermal degradation kinetics of various structurally different bisbenzoxazine systems (Shamim Rishwana 2013, 2015, 2016).

Structures of bisbenzoxazines derived from diphenol and diamine are presented in Figure 4. The superb molecular design flexibility of the polybisbenzoxazines allows the properties of the polymerized materials to be tailored in a wide range of properties for the specific requirements of individual application (Shamim Rishwana 2016).

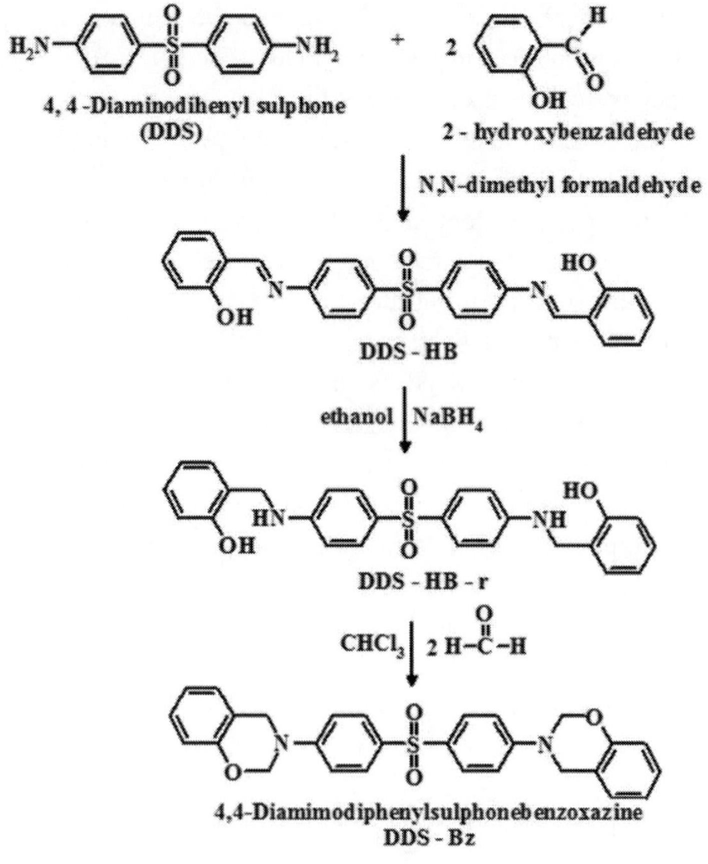

Figure 4. Structures of bisbenzoxazines from diphenol and diamine.

The technical and commercial importance of a polymer is in part based on the thermal stability of the material. Kinetic analysis of the thermal curing is essential for the processor and degradation of polymeric materials is essential for ascertaining their usage temperature. Thermal analyzers, Differential Scanning Calorimeter and Thermogravimetric analyzer, are extensively used to study the thermal properties of both the monomers and polymers. The kinetic parameters (apparent activation energy for thermal curing (Ea-C) and degradation (Ea-D), pre exponential factor (ln A) and reaction model (f(α)) of thermal curing and degradation give ample information regarding the thermal curing of the monomer and decomposition of the thermosets.

The compound DDS-Bz and PBS-a has been already synthesized and the characterization were reported by Agag et al. (Agag 2009) and Yangfang Liu (Liu 2010, 2011). In this present investigation, the authors wish to study both the curing aspect of these two bisbenzoxazines and also the thermal degradation behavior of the thermally polymerized materials. Most recent mathematical methods like Friedmann (Differential method), Flynn-Wall-Ozawa (FWO), Kissinger-Akahira-Sunose (KAS) and Vyazovkin (VYZ) – integral methods have been chosen to calculate the kinetic parameters and the results are presented and discussed in light of the existing ideas about curing and degradation mechanisms of bisbenzoxazine monomers and polymers.

EXPERIMENTAL

Materials

Preparation of 4,4-Diamino Diphenyl Sulphone Bisbenzoxazine (DDS-Bz)

An aromatic diamine, 4,4-diamino diphenyl sulphone (DDS) was employed as the starting material to form DDS-Bz. By following three step process (Lin 2010).

Step (I): 2-Hydroxybenzaldehyde (0.12 mol = 14.6 g) and DDS (0.06 mol = 14.9 g) were taken in a 250 mL round bottomed flask and dissolved in 100 mL of DMF with efficient stirring. The solution was kept at room temperature and stirred for three hours; the formed precipitate was filtered and dried for 24 h in an air oven kept at 80°C to obtain the first intermediate (DDS-HB) in powder form.

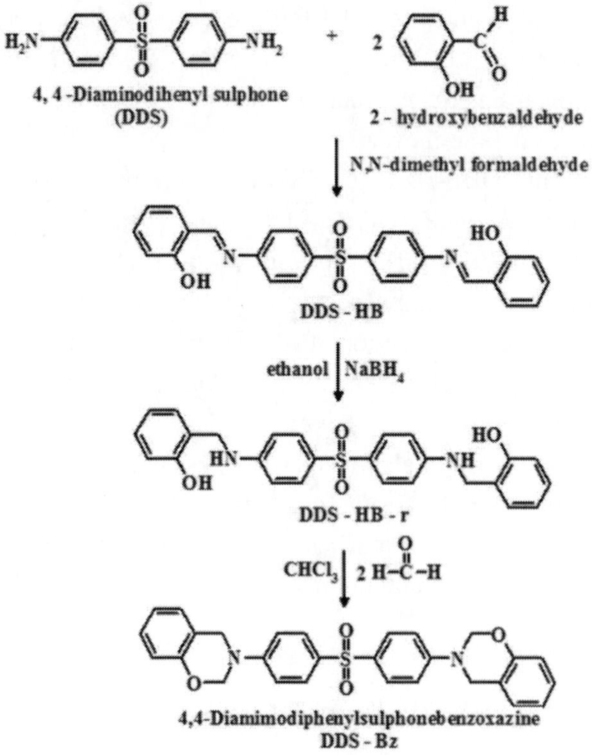

Figure 5. Preparation of 4,4-Diaminodiphenylsulphone bisbenzoxazine (DDS-Bz).

Step (II): Nitrogen was introduced into a reactor to remove the humidity for 30 minutes and a balloon with hydrogen was assembled on the reactor. The first intermediate (DDS-HB, 0.03 mol = 13.7 g) was dissolved in 30 mL of ethanol and taken in the reactor. Sodium borohydride (0.075 mol = 2.5 g) was divided into 3 batches and added to the reactant and the contents were kept at room temperature and stirred for

10 hours. The mixture was poured into water to precipitate the product formed. The precipitate was filtered and dried for 24 h in an air oven kept at 80°C to obtain the second intermediate (DDS-HB-r).

Step (III): In a 250 mL round flask the second intermediate (DDS-HB-r, 0.03 mol = 13.8 g) was dissolved in 85 mL of chloroform. Formaldehyde solution (0.06 mol = 4.6 mL) was added to the above solution in drop wise and the mixture was stirred for 4 hours at room temperature. The temperature was increased and the solution was refluxed for 5 hours. The solution was added into the ethanol solution (ethanol:water = 1:1) to precipitate the bisbenzoxazine (*DDS-Bz*). The three step process is presented in Figure 5.

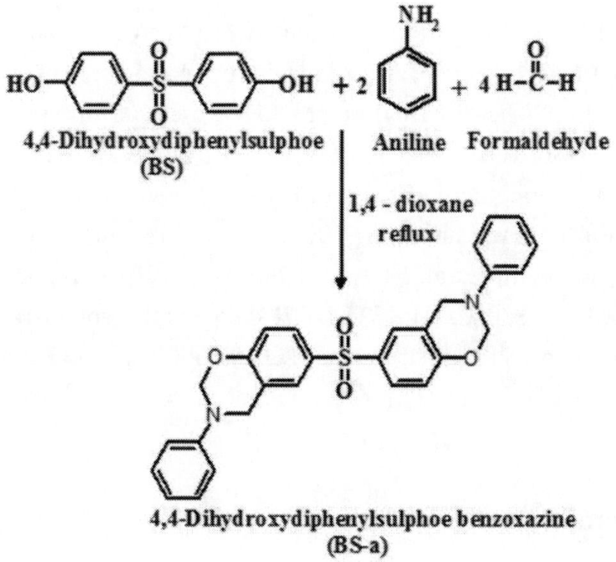

Figure 6. Preparation of 4,4-Dihydroxydiphenylsulphone bisbenzoxazine (BS-a).

Preparation of 6,6'-bis(3-phenyl-3,4-dihydro-2H-benzo[e][1,3]oxazinyl) Sulphone (BS-a)

In a 100 mL flask, aniline (0.06 mol = 5.9 g) was dissolved in 25 mL of 1,4-dioxane at room temperature. The solution was cooled in an ice bath, followed by the portion wise addition of paraformaldehyde (0.12 mol = 3.8 g) with stirring for 10 min. Then 4,4'-dihydroxydiphenylsulfone (BS)

(0.03 mol = 7.5 g) was added to this cold solution. The temperature was raised and refluxed for 24 h. After removing 1,4-dioxane under vacuum, the resulting crude product was purified by dissolving in 150 mL of diethyl ether and washing several times with 0.1N sodium hydroxide and finally two times with distilled water. After drying the diethyl ether solution with anhydrous sodium sulfate, evaporation of the ether lead to the product which was then dried under vacuum at 60°C for 24 h to afford solid (Kiskan 2009). The reaction is presented in Figure 6.

Methods

The FTIR spectra of the materials were recorded using Shimadzu (S8400) Fourier Transform Infra red spectrophotometer, Japan using KBr disc technique. The thermal analyses were performed using TA instruments DSC Q20 and TGA Q50. The DSC curves for the bisbenzoxazine monomers were recorded at different heating rates (β = 10, 20, 30°C min^{-1}) from ambient to 300°C. The TGA curves were recorded for the polymer samples at different heating rates (β = 10, 20, and 30°C min^{-1}) from 40 to 800°C. The TG-FTIR data for the polybisbenzoxazines were recorded in a TA Instruments TGA Q5000 V3.10 Build 258 at a heating rate of 10°C min^{-1}.

Kinetic Studies

The rate of solid-state reactions can be described as

$$\frac{d\alpha}{dt} = k(T) f(\alpha) \qquad (1)$$

where $d\alpha/dt$ is the rate of the reaction, $k(T)$ is the rate constant, $f(\alpha)$ is the reaction model. According to Arrhenius equation, the temperature-dependent rate constant, $k(T)$ is defined as

$$k(T) = A \exp\left(-\frac{Ea}{RT}\right) \quad (2)$$

where A is the pre-exponential factor, Ea is the apparent activation energy, R is the gas constant and T is the temperature.

The fractional degree of conversion (α) for the curing process is the expressed as:

$$\alpha = \Delta H_T / \Delta H_c \quad (3)$$

where, ΔH_T is the enthalpy of curing at particular temperature and ΔH_c is the total enthalpy of curing. The reaction rate ($d\alpha/dT$) for the curing process is given as:

$$d\alpha/dT = (dH/dT)/\Delta H_c \quad (4)$$

where dH/dT is the peak height at temperature T.

In the case of thermal degradation, the extent of conversion α, which is considered along with the mass of the reactant in the process, is formulated as follows

$$\alpha = (m_o - m)/(m_o - m_f) \quad (5)$$

where,
  m is the mass of the reactant at particular time/temperature
  m_o is the mass at the initial state
  m_f is the mass at the final state

The complete kinetic equations and other relevant details need for the calculation of the apparent activation energy for the thermal curing of the monomers (Ea-C) and the apparent activation energy for the thermal degradation of the polymers (Ea-D) are presented in the Appendix.

RESULTS AND DISCUSSION

FTIR Studies

The FTIR spectra recorded for the compounds DDS-Bz, and BS-a, are represented in Figure 7. The absence of a band at 2905 cm^{-1} in DDS-Bz indicates the –NH$_2$ group present in the amine have been used for the formation of oxazine rings. The presence of a band at 954 cm^{-1} (Agag 2009) is due to the N-C-O stretching vibrations. The symmetric and asymmetric stretching bands of the oxazine rings were noted at 1033 and 1226 cm^{-1} respectively. The C-N stretching band was noted at 1370 cm^{-1}. The formation of BS-a is due to the condensation reaction of the phenolic compounds with aniline and paraformaldehyde. The reduction in the intensity of the broad absorption peak typical for the –OH group (3450 cm^{-1}) in BS-a confirms the consumption of the –OH groups and leads to the formation of the oligomers. The band at 947 cm^{-1} is useful to recognize the oxazine ring structure. This band is due to the benzene ring mode of the benzene to which the oxazine is attached. Further the presence of the oxazine rings in the prepared compounds was supported by the appearance of new absorption bands for the oxazine asymmetric and symmetric stretching at 1229 and 1450 cm^{-1}, respectively (Liu, 2010-2011). The presence of C_6H_5-N group in BS-a was confirmed by an absorption band at 2900 cm^{-1}.

The FTIR spectra recorded for the polymers (PDDS-Bz and PBS-a) are also presented in Figure 7. The presence of an intense band at 3351 cm^{-1} indicates the formation of phenolic groups during the ring opening polymerization of the benzoxazines. When oxazine ring opens, the bands noted at 954 and 947 cm^{-1} disappears. It is important to note that disappearance of this mode is not the evidence of oxazine polymerization, but it is a simple indication that the oxazine ring gets opened. Further, the complete disappearance of the peaks specific for the oxazine group in the FTIR spectra of the polymers obtained by thermally polymerizing bisbenzoxazine monomers indicates the completion of the thermal polymerization.

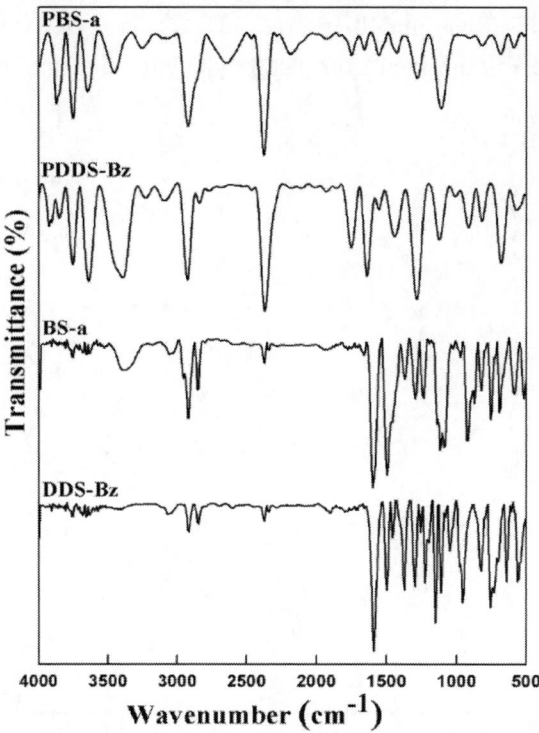

Figure 7. FTIR spectra of bisbenzoxazine monomers and their polymers.

The ladder structure resulting from the polymerization of DDS-Bz is shown in Figure 8. But the structure is much different from that of the structure noted for PBS-a. Here the poles of the ladder are joined by phenyl rings and this ladder is much flexible compared to that of the ladder resulting from the polymerization of BS-a. The PDDS-Bz shows intermolecular hydrogen bonding between the ladders through the favorably oriented phenolic –OH groups.

Polymerization of BS-a results in a ladder like structure (Figure 8) incorporating the benzene rings in the ladder. The appropriate orientation of the phenyl groups present in the nitrogen atom is a must to have intermolecular hydrogen bonding between the –OH groups present in the periphery of the ladders resulting in a structure that favors much thermal stability. Hence PBS-a is comparatively more stable than that of PDDS-Bz.

Shamim et al. (Shamim 2016) found similar observations for the bisbenzoxazines synthesized from *p*-phenylenediamine and quinol.

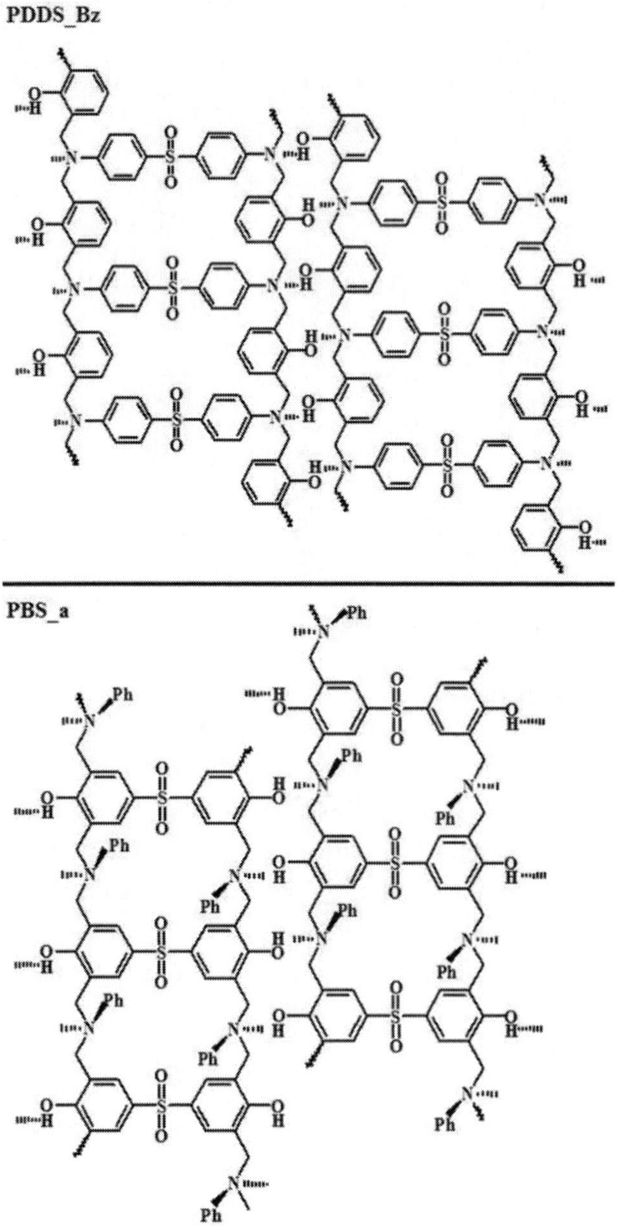

Figure 8. Structures of the polymers from bisbenzoxazines PDDS-Bz and PBS-a.

DSC Studies: Curing Behavior of Bisbenzoxazines

The DSC curves recorded at $\beta = 10°C$ min^{-1} for the compounds DDS-Bz and BS-a and the DSC curves for the two compounds at different heating rates (β = 10, 20, 30°C min^{-1}) are shown in Figure 9 and 10, respectively. The parameters obtained from the DSC curves, namely, onset (Ts), maximum (Tc max), end set (Te) and enthalpy of cure (ΔHc) reaction at multiple heating rates are compiled in Table 1 and discussed. The DDS-Bz shows a sharp and very high melting point in the range of 210 to 215°C when compared to BS-a which shows a broad and lower melting point at 80 to 90°C. Both the materials undergo polymerization and the polymerization exotherm maximum was seen at 247 (DDS-Bz) and 206°C (BS-a). The material DDS-Bz shows a sharp exotherm compared with BS-a and the processing window for BS-a is much wider (159-260°C = 101°C) compared with DDS-Bz (226-291°C = 65°C). The associated heat of enthalpy of cure reactions for DDS-Bz and BS-a are 321 and 212 J g^{-1}, respectively. The BS-a possesses the lowest value of enthalpy of curing as compared to that of the other compound. Compound DDS-Bz undergoes curing reactions at relatively higher temperature range (240-280°C) than BS-a. During thermal polymerization the formation of the Mannich base from DDS-Bz may be difficult. A comparison of these two bisbenzoxazines shows that the bisbenzoxazine formed from bisphenol results in lower curing temperature, lower heat release and with wider processing window.

Agag et al. (Agag 2009) for the first time synthesized the DDS based bisbenzoxazine. DDS-based bisbenzoxazine monomer was prepared from DDS as a weak aromatic diamine by solution method using nonpolar solvent such as xylenes at high temperature (150°C). The compound shows the exothermic peak in the range of 239°C which is much similar to the value for the compound synthesized by three step process of this investigation.

It can be observed that the exothermic peak shift to a higher temperature with higher heating rate. In these two systems, the heating rates show no marked effect on the total exothermicity of the curing reaction. The heat released during every degree rise in the temperature estimated from the area under the exothermic peak (ΔHc) and the curing window (Te-Ts) for DDS-Bz is found to be 4.5 - 5.0 J g^{-1}°C^{-1} and for BS-a is 1.8 - 2.1 J g^{-1}°C^{-1}. It is noticed that during the thermal curing reaction, DDS-Bz releases more amount of heat than that of BS-a. This suggests that BS-a is comparatively easy to thermally polymerize than DDS-Bz. These results suggest the nature and position of the oxazine group in a system influences the thermal curing behavior.

Table 1. DSC studies: The curing parameters for the different benzoxazines recorded at different heating rates

Sample	Temperature (°C)							ΔH$_C$ (J g^{-1})	ΔH$_C$/ Te– Ts (J g^{-1}°C^{-1})
	β	m.p.	Ts	Tc max	Te	Tc max - Ts	Te – Ts		
DDS-Bz	10	212	226	247	291	21	65	321	4.9
	20	213	234	262	310	28	76	340	4.5
	30	214	253	274	319	21	66	328	5.0
BS-a	10	81	159	206	260	47	101	212	2.1
	20	85	161	214	274	53	113	205	1.8
	30	88	168	221	281	53	113	205	1.8

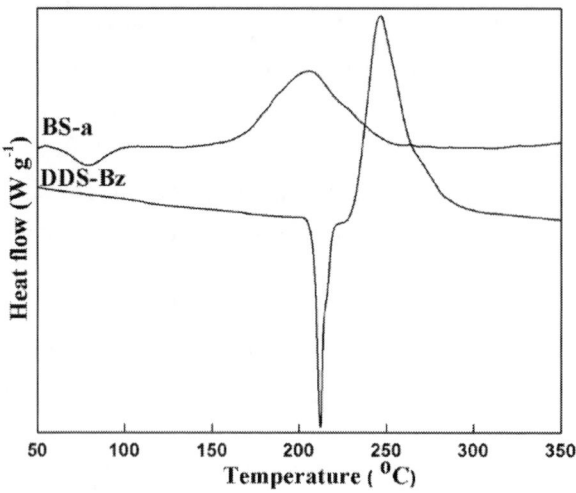

Figure 9. The DSC curves for DDS-Bz and BS-a recorded at β = 10°C min^{-1}.

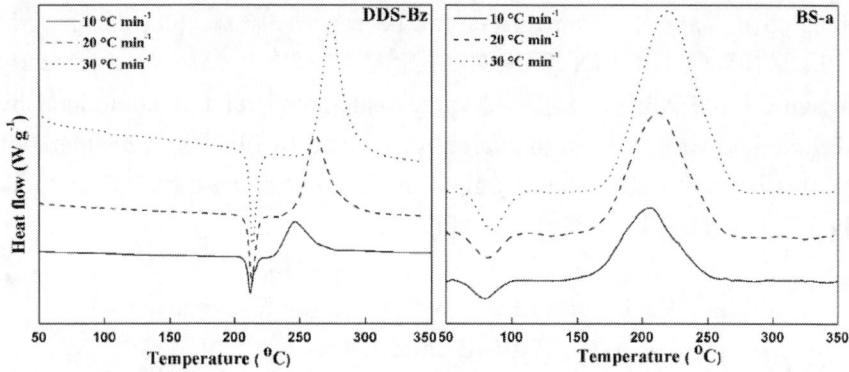

Figure 10. The DSC curves for DDS-Bz and BS-a recorded at different heating rates (β = 10, 20 and 30°C min^{-1}).

Curing Kinetics

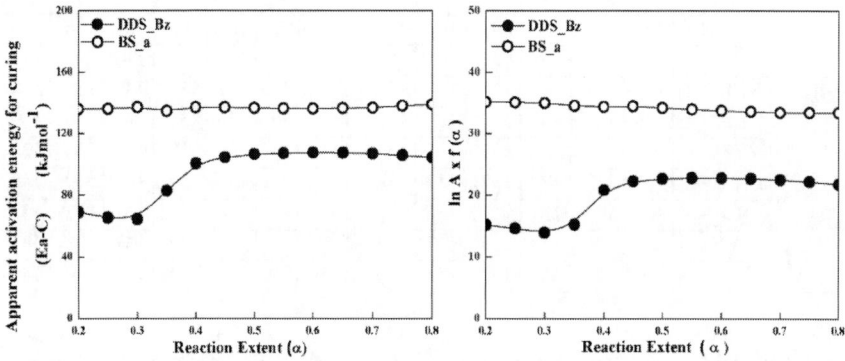

Figure 11. Application of A-VYZ method: Reaction extent (α) vs apparent activation energy for curing (Ea-C) and Reaction extent (α) vs ln A f(α) for the bisbenzoxazines (DDS-Bz and BS-a).

The different kinetic methods (FWO, C-FWO, KAS, C-KAS, VYZ, A-VYZ and FRD) were used to calculate the variation of the apparent activation energy for the curing (Ea-C) at different reaction extents (α) for the bisbenzoxazines DDS-Bz and BS-a. The plots between Ea-C and the reaction extent (α) values for the materials by the A-VYZ method is shown in Figure 11 and the values for the two monomers investigated are tabulated in Table 2. From the table one can easily see that the integral methods (FWO, C-FWO, KAS, C-KAS, VYZ and A-VYZ) lead to nearly the same Ea-C value. The Ea-C values calculated for this compound by FRD method various from the values calculated by other kinetics methods and this is due to the basic assumptions for developing the FRD method. The FRD method is a differential method.

Table 2. Variation of Ea-C with α for the bisbenzoxazines DDS-Bz and BS-a

Sample	Method/ α	Apparent activation energy for curing (Ea-C) kJ mol^{-1}											
		0.2	0.25	0.3	0.35	0.4	0.45	0.5	0.55	0.6	0.65	0.7	0.75	0.8
DDS-Bz	FWO	73	70	69	87	104	108	110	110	111	111	110	109	108
	KAS	68	65	64	82	100	104	106	107	107	107	107	106	104
	VYZ	69	65	65	83	101	104	107	107	108	108	107	106	105
	C-FWO	73	71	70	76	100	107	109	110	111	111	111	110	109
	C-KAS	68	66	64	71	96	103	105	107	107	107	107	106	105
	A-VYZ	69	66	65	80	100	104	107	107	108	108	107	106	105
	FRD	71	69	69	93	105	105	112	108	109	115	116	109	112
BS-a	FWO	137	137	138	135	138	138	137	137	137	137	138	139	140
	KAS	136	136	137	135	137	137	137	136	136	136	137	138	139
	VYZ	136	136	137	135	137	137	137	136	137	137	137	138	139
	C-FWO	137	137	137	137	137	138	138	137	137	137	138	139	140
	C-KAS	136	136	137	136	136	137	137	136	136	136	137	137	138
	A-VYZ	136	136	137	135	137	137	137	136	137	137	137	138	139
	FRD	134	139	139	132	138	139	133	134	136	135	140	140	140

The Ea-C value for DDS-Bz was observed to increase gradually with increasing extent of reaction (α). In this system, Ea-C calculated by the A-VYZ method for the DDS-Bz polymerization increases from 69 to 105 kJ mol^{-1} as the extent of the reaction (α) value increases from 0.2 to 0.65. Slight decrease in activation energy is noted after the α value of 0.65. The rate of increase in Ea-C value was different. Initially, the concentration of the monomer was high and hence, the polymerization reaction was easier and required low energy for the reaction. As the reaction proceeds, viscosity increases and the availability of the monomer for the polymerization decreases. The increase in the viscosity and the dearth in polymerizable groups made the crosslinking reaction difficult. As the temperature increased linearly in the non isothermal experiments, the chain mobility increased and the chemical reactions were reactivated. This results in the pattern noted for the variation of Ea-C with respect to change in extent of reaction (α).

The recent work done by Ishida et al. (Ishida 1995) examined the curing kinetics of benzoxazine resin (Bisphenol-A benzoxazine (BAB)) with and without catalysts by using both isothermal and non isothermal differential scanning calorimetry (Jubsilp 2006). They reported that the curing of benzoxazine precursor was an auto catalyzed reaction prior to diffusion control stage. The apparent activation energy by KAS and FWO method of the curing process was found to be about 102-116 kJ mol^{-1} in an uncatalysed system and 99 - 107 kJ mol^{-1} in a catalyzed system with an overall order of reaction of about 2. The phenol moiety of the ring opened benzoxazine monomers was reported to have a catalytic effect on the curing reaction i.e., reducing the reaction induction time and increasing rate. The isothermal curing process of the polybenzoxazine precursor involves an autocatalytic type curing mechanism. Similarly in BS-a, the polymerization is an autocatalytic mechanism because of the presence of the phenolic groups in the oligomer.

Liu et al. (Liu 2010) by using the dynamic polymerization mode calculated the activation energy values for the thermal polymerization of BS-a and reported 176.8 kJ mol^{-1} by KAS method and 175.7 kJ mol^{-1} by FWO method. In isothermal polymerization mode, the activation energy values of BS-a is 140.7 kJ mol^{-1}. Similar results have been observed for BS-a, where the apparent activation energy for the thermal curing vary from 136 - 139 kJ mol^{-1}. The Ea-C for the thermal curing of BS-a is not affected much with increasing α values. The consistency in the Ea-C values may be attributed to the occurrence of similar reaction with similar ease throughout the conversion range.

The higher apparent activation energy for the thermal curing of the monomer bisbenzoxazine BS-a when compared to that of bisphenol-A bisbenzoxazine (BAB)—may be ascribed to the structural difference between BS-a and BAB monomers, wherein BS-a possesses the rigid sulfone moiety, which is an electron withdrawing group But BAB is having the flexible isopropyl electron donating group. This difference will play a role in the energy that has been required for the opening the benzoxazine ring. Compared to DDS-Bz, BS-a shows higher Ea-C for curing reaction. This may be due to the stability of the ring system present in the BS-a. Owing to the ring stability, the energy needed for the ring opening polymerization is much higher for BS-a when compared to DDS-Bz

From the intercepts of the linear lines obtained by the application of FWO, KAS, C-FWO and C-KAS methods to the DSC curves recorded at different heating rates, the parameter ln A f(α) can be calculated. From the previous inferences, this parameter for each conversion calculated from the C-FWO and C-KAS methods are more reliable since the systematic errors in the numerical approximations of Doyle and Coats-Redfern temperature integral does not follow in those equations. However, it is advisable to take the intercept of the last iteration of the C-FWO or C-KAS equation to estimate the parameter ln A f(α).

$$A = \left(EG(\alpha)/R \right) \exp(INT) \tag{6}$$

$$A = \left(RG(\alpha)/E \right) \exp(INT + 5.330) \tag{7}$$

The INT in equations (6) and (7) are respectively the intercepts of the last iterations obtained from C-KAS and C-FWO method. The parameter ln A f(α) obtained are plotted in Figure 8. Much similar to the Ea-C values, the pre exponential factor along with the reaction model is also a parameter of temperature; this parameter shows significant changes with change in reaction extent.

Kessler and white et al. (Kessler & white 2002) studied the curing kinetics of ring opening metathesis polymerization of dicyclopentadiene. From this result it was found that, the plots of ln A versus α and Ea-C versus α appear very similar. This correspondence is due to the isokinetic relationship (Friedman 1964, Vyazovkin 1997) or the kinetic compensation effect, (Pielichowski 2000) which suggests that the value of ln A varies with the variation in the value of Ea-C. Such relationship has been observed in the curing and decomposition of numerous other polymer systems (Cooney 1984, Montserrat 1996). According to the statistical procedure proposed by Vyazovkin and Wight (Vyazovkin & Wight 2000) the realistic confidence intervals for the Ea-C determined are estimated for both DDS-Bz and BS-a and are presented in Figure 12. The resulting averaged relative error in the Ea-C was found to be approximately 25 %.

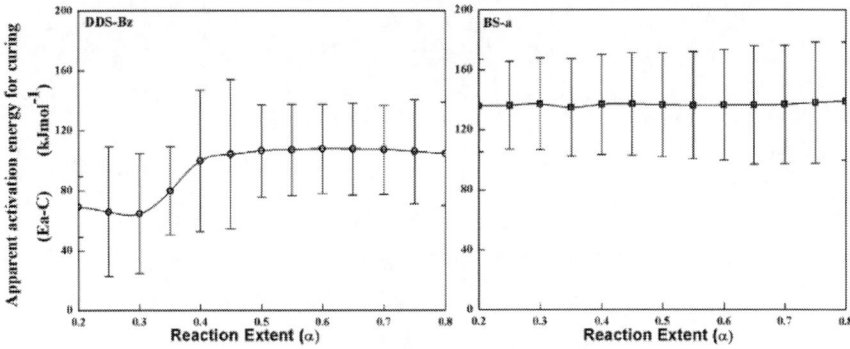

Figure 12. Apparent activation energy for curing (Ea-C) and Fisher confidence intervals for the bisbenzoxazines as estimated by A-VYZ method.

TG and DTG Studies

Thermogravimetric Analysis (TGA) is one of the commonly used techniques for the evaluation of thermal stability of materials and also gives precise idea regarding the degradation temperatures of various materials. The TG curves and differential thermogravimetric (DTG) curves for thermally cured materials, PDDS-Bz and PBS-a recorded at 10^0C min^{-1} and at various heating rate (β = 10, 20 and 30^0C min^{-1}) in nitrogen atmosphere are shown in Figure 13 and 14 respectively. All the thermograms are shifted to higher temperatures with increasing heating rates. The parameters obtained from the TG and DTG curves, namely degradation onset temperature (Ti), temperature at which occurrence of 5 %, 10 % and 25 % mass losses (T5, T10 and T25), temperature at which the degradation is maximum for the various degradation stages (Td max) and end set temperature (Te) of degradation are tabulated in Table 3. The detailed observations of the TG data for both the samples obtained at a heating rate of 10°C min^{-1} are discussed below.

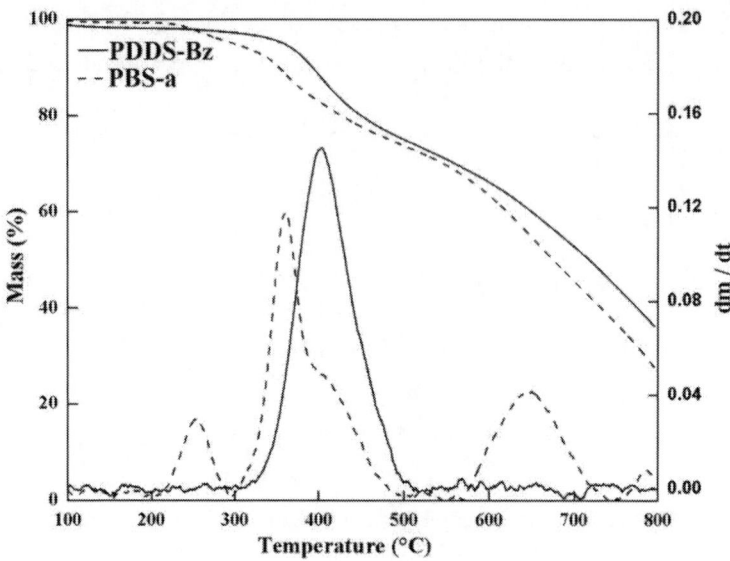

Figure 13. The TG and DTG curves for PDDS-Bz and PBS-a recorded at a heating rate of 10°C min^{-1}.

Table 3. TG studies: The degradation parameters for the two polybisbenzoxazines recorded at the different heating rates (β = 10, 20 and 30°C min^{-1})

Sample	Temperature (°C)						Td max (°C)		
	β	Ti	T5	T10	T25	Te	m1	m2	m3
PDDS-Bz	10	320	350	392	506	513		400	
	20	324	369	405	498	540		418	
	30	336	370	409	499	546		426	
PBS-a	10	212	298	353	483	745	251	358	646
	20	230	307	363	495	752	285	376	665
	30	244	314	369	495	759	275	381	673

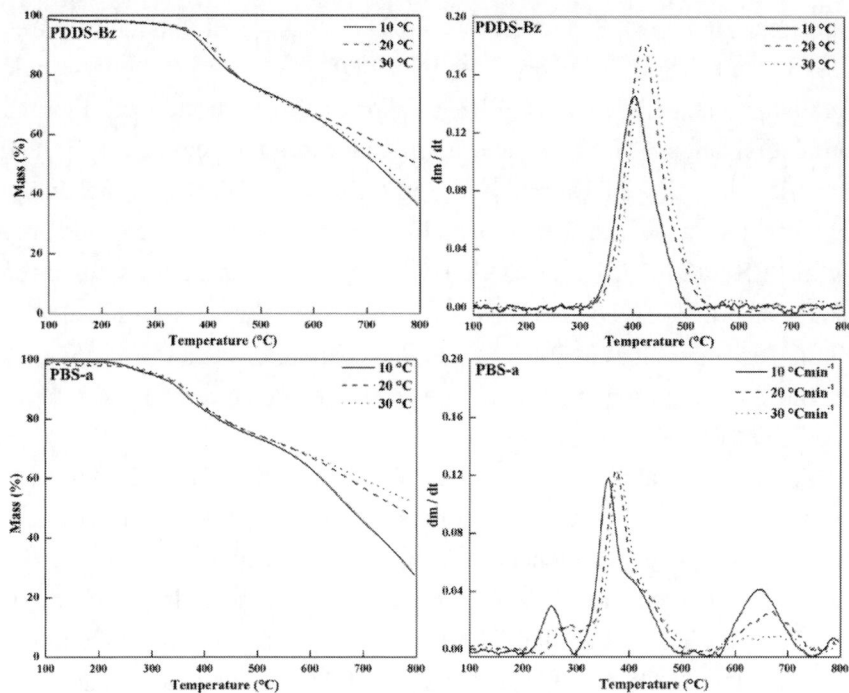

Figure 14. The TG and DTG curves for PDDS-Bz and PBS-a recorded at different heating rates (β = 10, 20 and 30°C min^{-1}).

The benzoxazine ring is having distorted semi chair structure. This molecular conformation results in high strain, which favours the six membered ring molecules to undergo ring opening polymerization (Vijayakumar 2013, Wang 2000). Ionic mechanism and attempts to study such mechanism have been made regarding the ring-opening polymerization of benzoxazine. The formation of iminium ion during the polymerization is the most probable route for the polymerization of these materials. Still the polymerization mechanism of benzoxazine resin has been elusive. The mechanistic complexity existing in the polymerization of bisbenzoxazine may be attributed to the different sites of the benzene rings having varying degrees of reactivity.

The thermally cured DDS-Bz and BS-a show the onset degradation temperature at 320 and 212°C, the thermal degradation maximum at 400 and 358°C and the thermal degradation ends at 513 and 745°C, respectively (Figure 13 and 14). From DTG curves of thermally cured DDS-Bz and BS-a, it is obvious that PDDS-Bz nearly shows single degradation stage, whereas PBS-a shows three prominent thermal degradation stages. The 5, 10 and 25 % mass loss temperatures (T5, T10 and T25) for the PDDS-Bz are 350, 392 and 506°C, respectively. However, the 5, 10 and 25 % mass loss temperatures were considerably low in PBS-a (298, 353 and 483°C). Agag et al. (Agag 2009) studied the thermal stability of PDDS-Bz and they reported 5 % mass loss at 324°C and the 10 % mass loss at 368°C. Also, the char yield was 58 %.

The difference noted may be due to the temperature and the time at which DDS-Bz was polymerized.

Further thermally cured material from the DDS-Bz degrades in a narrow temperature range when compared to the polymer derived from BS-a. Although the polymer PBS-a starts to degrade at an earlier temperature, the material degrades slowly and the degradation proceeds upto 745°C, whereas the thermal degradation goes to near completion at 513°C for the PDDS-Bz. PDDS-Bz showed 2 % less average char value (41 %) compared to PBS-a (43 %). From this study, one can easily infer that the structure resulting from the polymerization of DDS-Bz is definitely different from the molecular structure resulting from the thermal polymerization of BS-a.

Liu et al. (Liu 2011) studied the thermal degradation behavior and kinetics of polybenzoxazine based bisphenol-S and aniline, they reported degradation of the compound in two different atmospheres. The primary mass loss behavior in the degradation is most easily identified and the degradation process in nitrogen and air can be roughly divided into two mass loss stages. In the first stage, it can be evaluated that the mass loss amounts to about 20 % in both nitrogen and air. In the second stage, the mass loss behavior in nitrogen is significantly different from that in air. Though trace of C_6H_6 was released over the higher temperature range, about 65 % residues remained at 800°C in nitrogen. In our system the material PBS-a shows the average char residue of 43 % at 800°C this may be due to the variation in the polymerization temperature and time. Hence it is clear that the selection of the curing temperature of bisbenzoxazine is playing a vital role in determining the thermal stability of the resulting polymer.

Vijayakumar et al. (Vijayakumar 2013) studied synthesis, polymerization and thermal stability of structurally diverse bisbenzoxazines and reported that the spirobiindane unit containing polybenzoxazine (PSBIB) when cured at higher temperature is found to lose its thermal stability and is reflected in the decrease in the degradation temperature (227°C) and final char value (22 %).

Degradation Studies

The apparent activation energies for degradation (Ea-D) for the thermal degradation at different reaction extents (α) were calculated for the temperature region 300 to 450°C, where the maximum degradation takes place. The Ea-D values for the degradation of the polymers were determined by different model free kinetic methods and are presented in Table 4. The plots between Ea-D versus α and ln A f(α) versus α of thermally cured materials by A-VYZ are shown in Figure 15. The Ea-D obtained by using A-VYZ method for thermally cured materials is considered for the discussion.

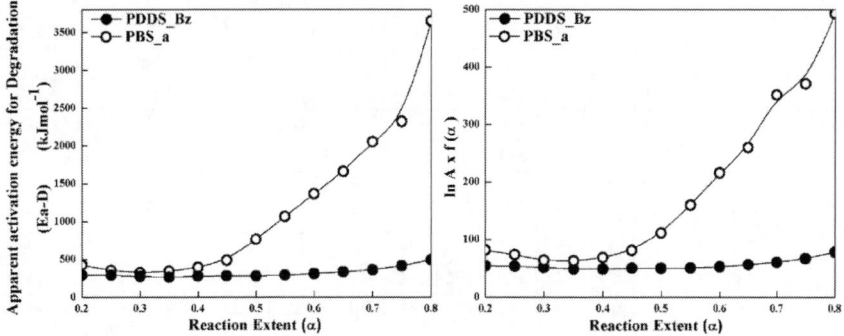

Figure 15. Plots of reaction extent (α) vs apparent activation energy for degradation (Ea-D) and reaction extent (α) vs ln A f(α) for the polymers PDDS-Bz and PBS-a (A-VYZ method).

The apparent activation energy for the degradation of PDDS-Bz and PBS-a varies from 293 to 494 kJ mol^{-1} and 424 to 3650 kJ mol^{-1} respectively. Initially the apparent activation energy for degradation (Ea-D) for PDDS-Bz gradually decreases upto the reaction extent level of 0.35. After the reaction extent of 0.35, the Ea-D value increased steadily for higher values of α. The initial decrease in Ea-D values was attributed to the existence of weak points in the polymeric chains whereas the higher Ea-D values at the later stages of degradation was associated with high degrees of random scission of the main chain. The initial lower value of the apparent activation energy is mostly likely associated with the initiation process that occurs at these weak links. As these weak links are consumed, the limiting step of degradation shifts toward the degradation initiated by random scission, which typically has greater activation energy (Pitchaimari 2014).

Table 4. Ea-Ds for thermal degradation at different αs for PDDS-Bz and PBS-a

Sample	Method / α	Apparent activation energy for degradation (Ea-D) kJ/mol												
		0.2	0.25	0.3	0.35	0.4	0.45	0.5	0.55	0.6	0.65	0.7	0.75	0.8
PDDS-Bz	FWO	286	285	270	261	274	277	276	287	305	328	353	405	482
	KAS	290	289	273	263	278	280	279	290	310	334	360	414	496
	VYZ	290	289	274	263	278	281	279	290	310	334	361	414	496
	C-FWO	286	285	278	265	267	277	277	281	296	316	340	376	437
	C-KAS	290	289	281	268	270	280	280	284	300	322	347	386	453
	A-VYZ	293	289	274	264	277	281	279	290	310	333	360	413	494
	FRD	271	275	262	269	261	252	291	254	341	370	319	357	506
PBS-a	FWO	409	342	315	334	383	472	741	1027	1311	1589	1960	2217	3486
	KAS	420	349	321	341	392	486	769	1069	1368	1661	2050	2321	3656
	VYZ	420	349	321	341	392	486	769	1069	1368	1661	2050	2322	3656
	C-FWO	409	370	328	325	355	418	526	798	1134	1384	1934	2058	2525
	C-KAS	420	381	334	331	365	436	608	882	1206	1472	2023	2165	2922
	A-VYZ	424	351	322	340	391	484	763	1064	1365	1658	2050	2320	3650
	FRD	399	469	333	212	499	438	781	1082	1373	1571	1983	2131	3423

Initially for PBS-a, the apparent activation energy for degradation value (Ea-D) increases with increasing reaction extent values (α). After that the Ea-D values shows a sudden increase of Ea-D for further α values.

This may be due to the difference in the structure of the network formed and the way in which the degradation is initiated and propagated. The very high values of Ea-D calculated at the higher extents of degradation explain the resistance shown by the structure of the material under investigation which consequently increases the final char of the system.

Liu et al. (Liu 2011) studied the activation energies for thermal degradation of PBS-a for each degradation stages separately. They calculated based on the conversion values ranged from 0.1 to 0.7 in the first degradation stage and these values show that the presence of oxygen enhances the reaction activation energy. To best fit the experiment results, the most frequently used mechanism based kinetic models were evaluated for the thermal degradation kinetics of PBS-a in nitrogen and in air. The conversion values ranged from 0.1 to 0.45 in the first stage were used to test various mechanism functions.

Krongauz (Krongauz 2010) reported similar results for the crosslink density dependence of polymer degradation kinetics in photo crosslinked acrylates. The rotation of polymer segments is consistent with the low apparent activation energy of polyacrylate degradation. As the crosslinking density increases, an increase in the activation energy is noted. Change from degradation due to rotation of polymer segments at low temperatures to degradation through direct bond scission at higher temperature may also explain the increase in apparent activation energy with temperature.

The pre exponential factor (ln A) along with reaction model f(α) was estimated and the plot is shown in Figure 15. The ln A f(α) value is lower for the compound PDDS-Bz when compare to the PBS-a. The value varies from 54 to 78 s^{-1} and 81 to 493 s^{-1} for PDDS-Bz and PBS-a respectively. The trend in the variation of ln A f(α) is much similar to the trend obtained for the plot between α and Ea-D.

A statistical procedure is used for estimating the confidence intervals for the activation energy evaluated by the use of an advanced isoconversional method. This statistical procedure was proposed by Vyazovkin and Wight (Vyazovkin & Wight 2000). The realistic confidence intervals for the activation energy determined are estimated, the estimated activation energy and the Fisher confidence intervals for the three heating rates (β = 10, 20 and 30°C min⁻¹) for the samples PDDS-Bz and PBS-a are presented in Figure 16. These errors may appear somewhat greater than one would wish for, but they are significantly smaller than the actual uncertainty in the activation energies estimated by fitting various reaction models to single heating rate data (Vyazovkin & Wight 1998). The resulting averaged relative error in the activation energy was found to be approximately 25 %.

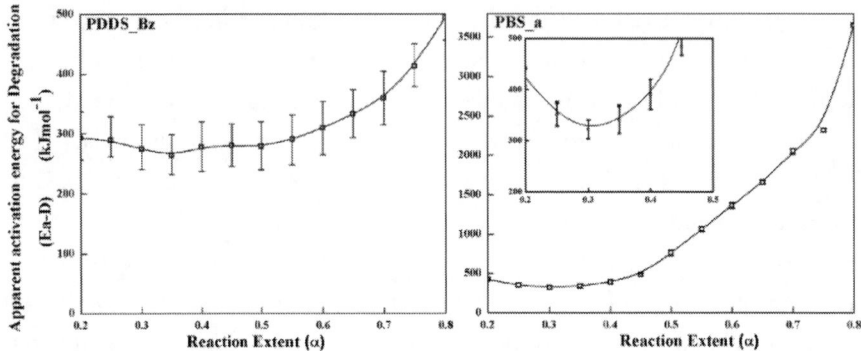

Figure 16. Apparent activation energy for degradation and Fisher confidence intervals for the compounds (PDDS-Bz and PBS-a) at heating rates (β = 10, 20, and 30°C min⁻¹) estimated by A-VYZ method.

The thermal lifetime for the compounds (PDDS-Bz and PBS-a) have been estimated from equation (8) for the reaction extent (or failure of conversion, in this case) of 20 % (α = 0.2).

$$\ln t_f = \frac{E}{RT_\alpha} + \ln\left[\frac{Ep(x)}{\beta R}\right] \tag{8}$$

where, t is the thermal lifetime of the materials, p(x) is the temperature integral of first order, β is the heating program in °C min⁻¹ and T_α is the temperature at the conversion α where the required set of properties of the material can be maintained.

The optimum working temperature at which the degradation of the compound take place at the heating rate of 10°C min⁻¹ has been estimated for different working hours (Figure 17). The working temperature of the compound reduces exponentially with the increase in required thermal lifetime. The compound PDDS-Bz starts to fail after 20,000 h provided the working temperature is 254°C but for the PBS-a it was found to be 238°C. The systems fail in 1 h when the working temperature is 326°C for PDDS-Bz and 317°C for PBS-a.

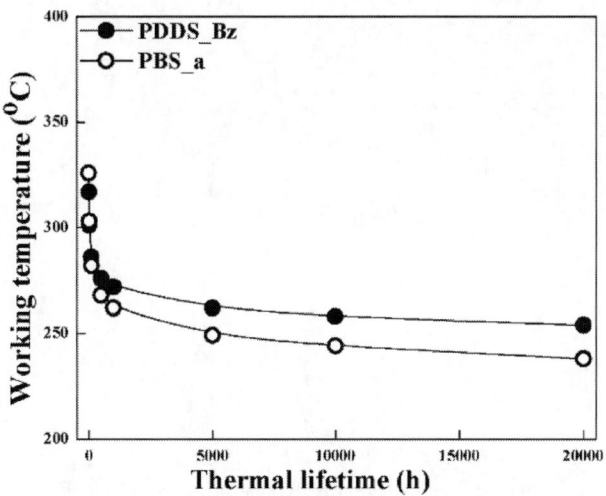

Figure 17. Working temperature versus thermal lifetime for degradation process at failure of conversion ($\alpha = 0.2$).

TG-FTIR Studies

The combination technique TG-FTIR provides information on neither the exact structure of all the decomposition products nor the separation of different decomposition products coming off at the same time. The processes of thermal degradation of various polybenzoxazines were studied and the degradation mechanisms were reported in the literature (Low & Ishida 1998, Low & Ishida 1999). The structural effects of polymeric compounds on thermal degradation were investigated systematically.

The TG-FTIR studies of polybenzoxazines prepared using sulphone based diamine and diphenol have been carried out and the results are presented in Figure 18. The FTIR shown in the Figure 18 are for the mixture of products resulting from the thermal degradation of the chosen compound.

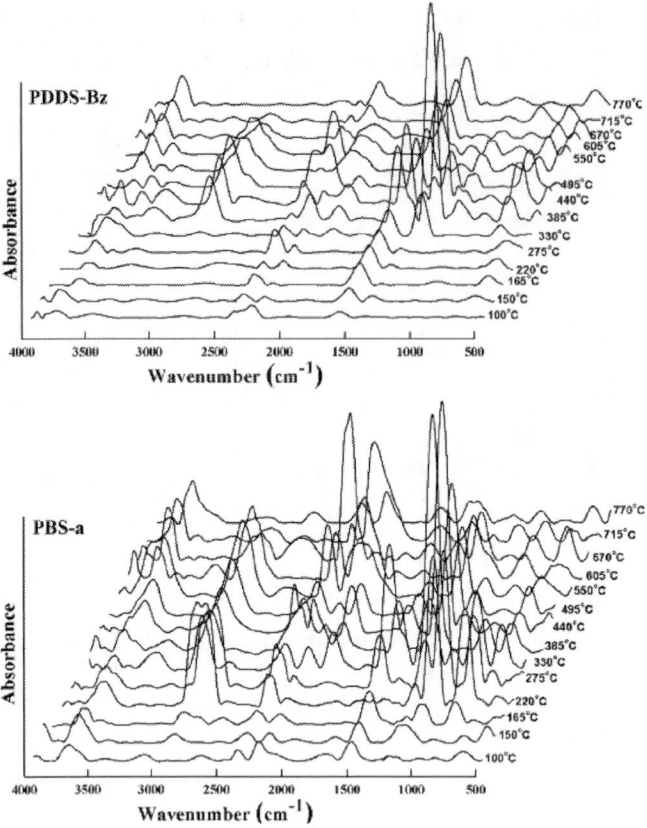

Figure 18. The TG-FTIR curves of PDDS-Bz and PBS-a.

The bond dissociation energy is an important factor affecting the degradation mechanism of the compound and also the Ea-D which is related to the bond dissociation energies of various linkages in the compounds. The C-S bond dissociation energy (bond energy-272 kJ mol^{-1}) is lower than the other bonds dissociation energies. Hence it is reasonable to expect the initial degradation take place at C-S bond. The next likely cleavage point is the aromatic CCH$_2$-N bond (bond energy-304 kJ mol^{-1}). A Schiff base (C=N) is formed when the Mannich base is cleaved. So, the next likely cleavage point is the C-C bond as the C-C bond (bond energy-

364 kJ mol^{-1}) has much lower bond energy than a C-N bond (bond energy-429 kJ mol^{-1}).

In the compound PBS-a at the initial stages of thermal degradation, aromatic amine and phenol were the volatile products. Large quantities of the volatiles are formed around 330 to 495°C. In this temperature range, the cleavage of C-S bond takes place. Based on the TG-FTIR evolved gas analyses the probable thermal degradation products from PDDS-Bz and PBS-a have been proposed (Figure 19).

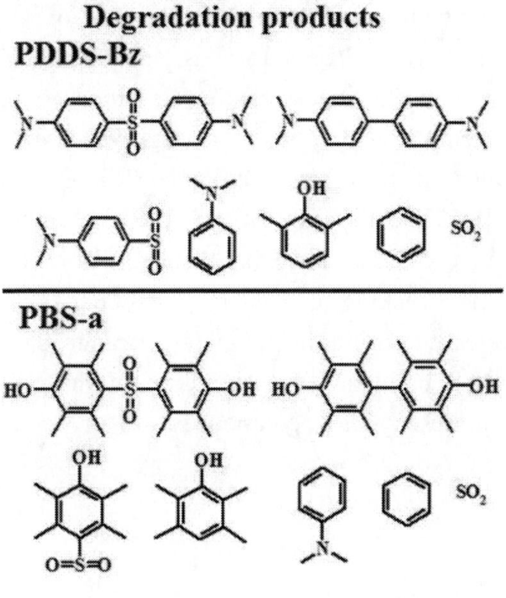

Figure 19. The possible degradation products from PDDS-Bz and PBS-a.

From Figure 19 one can conclude that aromatic amine, substituted aromatic amine and biphenyl compounds were the major degradation products at the degradation stage where the rate of degradation is found to be the maximum. The Schiff base was also detected as a degradation product from the bisbenzoxazine at around 400°C. The appearance of bands around 3068, 3486 and 3785 cm^{-1} were significant. This is mainly attributed to the evolution of aromatic diamine derivatives. This temperature corresponds to the temperature of the maximum rate of mass loss in the degradation curves of the compound PDDS-Bz.

The formation of phenols from PBS-a, needs a minimum of four bond cleavages (Figure 19). At around 358°C, significant bands at 2898, 2397 and 3660 cm^{-1} were detected from degrading PBS-a. These are due to the free OH group of phenol and it is accompanied by a band at 1179 cm^{-1}, which is due to the C-O bond of phenol or substituted phenols. Hence primary or the initial degradation of PBS-a is the formation of volatile aromatic amines followed by the evolution of phenolic compounds and this represents the temperature of the maximum rate of mass loss in the thermogravimetric curves of PBS-a. Thus, the degradation observed at higher temperatures in the curve can be assigned to the evolution of phenolic compounds from the diphenol based polybisbenzoxazines.

The appearance of SO_2 bands around 1361 and 1346 cm^{-1} were significant for both PDDS-Bz and PBS-a at the temperature region of 400 and 358°C respectively. This shows that the degradation of the compounds dependence on the breaking up of the C-S bond. Wide varieties of compounds are released during this temperature range. This shows that the cleavage of the C-S bond and the evolution of the SO_2 determine the thermal stability of these types of compounds.

CONCLUSION

The two compounds DDS-Bz and PBS-a were prepared and polymerized. The structures of the synthesized and polymerized compounds were confirmed by FTIR studies. The thermal curing of the bisbenzoxazine monomers and the thermal stabilities of the polybisbenzoxazines were studied in detail using DSC and TGA respectively. It was found that the orientation of the polymerizable group is having the significant influence on the thermal curing and degradation parameters. The kinetics of curing and degradation of these materials were studied using different advanced isoconversional kinetic methods. The ln A f(α) parameters both for the thermal curing of the monomers and thermal degradation of the polybisbenzoxazines have also been estimated. The thermal life time of the polymers were predicted from the kinetic data obtained. The evolved gases during the thermal degradation were analyzed using TG-FTIR technique. Apart from SO_2, aromatic amines, phenols and their derivatives, were formed during the thermal degradation of the cured materials.

CONFLICT OF INTEREST

On behalf of all authors, the corresponding author states that there is no conflict of interest.

REFERENCES

Agag T, Jin L, and Ishida H, 2009 "A new synthetic approach for difficult benzoxazines: Preparation and polymerization of 4,4'-diaminodiphenyl sulfone-based benzoxazine monomer." *Polymer* 50: 5940–5944. doi:10.1016/j.polymer.2009.06.038.

Burke, WJ 1949, "3, 4-Dihydro-1, 3,2H-Benzoxazines. Reaction of p-Substituted Phenols with N, N-Dimethylolamines." *Journal of American Chemical Society*, 71: 609-612.

Burke, WJ, Kolbezen, MJ, and Stephens, CW 1952, "Condensation of Naphthols with Formaldehyde and Primary Amines." *Journal of American Chemical Society*, 74: 3601-3605.

Cooney JD, Day M, and Wiles DM. 1984. "Kinetic and thermogravimetric analysis of the thermal oxidative degradation of flame - retardant polyesters." *Journal of Applied Polymer Science* 29: 911–923. https://doi.org/10.1002/app.1984.070290319.

Dodiuk, H, and Goodman, SH 2014, *"Introduction. Handbook of Thermoset Plastics"*, 1–12. DOI:10.1016/b978-1-4557-3107-7.00001-4.

Farjas J and Roura P. 2011. "Isoconversional analysis of solid state transformations." *Journal Thermal Analysis.* 105: 757-766.

Flory, PJ 1953, *Principles of Polymer Chemistry,* Cornell University, Ithaca, New York.

Flynn JH and Wall LA. 1996. "General treatment of thermogravimetry of polymers." *Journal of Research of the National Bureau of Standards. Section A, Physics and Chemistry.* 70: 487-524. DOI: 10.6028/jres.070A.043.

Friedman HL. 1964. "Kinetics of thermal degradation of char-forming plastics from thermogravimetry. Application to a phenolic plastic." *Journal of Polymer Science Part C: Polymer Symposia.* 6: 183-195. https://doi.org/10.1002/ polc.5070060121.

Hergenrother, PM 2003, "The use, design, synthesis and properties of high performance/high temperature polymers: an overview." *High Performance Polymers,* 15: 3-45.

Higginbottom, HP 1985, "Polymerizable compositions comprising polyamines and poly (dihydrobenzoxazines)." U.S. Patent. 4,501,864.

Holly, FW and Cope, AC 1944, "Condensation Products of Aldehydes and Ketones with o-Aminobenzyl Alcohol and o-Hydroxybenzylamine." *Journal of American Chemical Society,* 66: 1875-1879.

Ishida, H 1996, *Process for preparation of benzoxazine compounds in solventless systems.* US patent 5,543,516.

Ishida, H and Allen, DJ 1996, "Physical and mechanical characterization of near-zero shrinkage polybenzoxazines." *Journal of Polymer Science, Part B: Polymer Physics,* 34: 1019-1030.

Ishida, H and Lee, YH 2001, "Study of hydrogen bonding and thermal properties of polybenzoxazine and poly-(epsilon-caprolactone) blends." *Journal of Polymer Science Part B: Polymer Physics,* 39: 736–749.

Ishida, H and Rodriguez, Y 1995, "Curing kinetics of a new benzoxazine-based phenolic resin by differential scanning calorimetry." *Polymer,* 36: 3151-3158.

Ishida, H and Sanders, DP 2000, "Improved thermal and mechanical properties of polybenzoxazines based on alkyl-substituted aromatic amines." *Journal of Polymer Science Part B: Polymer Physics*, 38: 3289–3301.

Jubsilp C, Damrongsakkul S, Takeichi T and Rimdusit S. 2006. "Curing kinetics of arylamine-based polyfunctional benzoxazine resins by dynamic differential scanning calorimetry." *Thermochimica Acta*, 447: 131-140. doi.org/10.1016/j.tca.2006.05.008.

Kessler MR and White SR. 2002. "Cure Kinetics of the Ring-Opening Metathesis Polymerization of Dicyclopentadiene." *Journal of Polymer Science. Part A Polymer Chemistry*. 40: 2373–2383. https://doi.org/10.1002/pola.10317.

Kim HD and Ishida H 2001. "Study on the chemical stability of benzoxazine-based phenolic resins in carboxylic acids." *Journal of Applied Polymer Science*, 79: 1207-1219.

Kiskan B, Aydogan B, and Yagci Y. 2009. "Synthesis, characterization, and thermally activated curing of oligosiloxanes containing benzoxazine moieties in the main chain." *Journal of Polymer Science Part A Polymer Chemistry* 47: 804–811. https://doi.org/10.1002/pola.23197.

Kissinger HE. 1957. "Reaction kinetics in differential thermal analysis." *Analytical Chemistry*. 11: 1702–1706. https://doi.org/10.1021/ac60131a045.

Knop, A, Pilato, LA 1985, 'Phenolic Resin', Springer-Veriag, New York.

Krongauz VV. 2010. "Crosslink density dependence of polymer degradation kinetics: Photocrosslinked acrylates." *Thermochimica Acta* 503: 70-84. https://doi.org/10.1016/j.tca.2010.03.011.

Liang JZ, Wang JZ, Tsui, GCP and Tang CY. 2015." Thermal decomposition kinetics of polypropylene composites filled with graphene nanoplatelets." *Polymer Testing*. 48: 97-103. https://doi.org/10.1016/j.polymertesting.2015.09.015.

Lin CH, Taso YR, Sie JW and Lee HH. 2010. *Processes of synthesis of aromatic amine-based benzoxazine resins.* US Patent 7781561.

Liu Y, Yue Z and Gao J. 2010. "Synthesis, characterization, and thermally activated polymerization behavior of bisphenol-S/aniline based

benzoxazine." *Polymer* 51: 3722-3729. doi.org/10.1016/j.polymer.
2010.06.009.

Liu Y, Yue Z, Li Z and Liu Z. 2011. "Thermal degradation behavior and kinetics of polybenzoxazine based on bisphenol-S and aniline." *Thermochim. Acta* 523: 170–175. doi:10.1016/j.tca.2011.05.020.

Low HY and Ishida H. 1998. "Mechanistic study on the thermal decomposition of polybenzoxazines: effects of aliphatic amines", *Journal of Polymer Science Part B Polymer Physics* 36: 1935–1946. https://doi.org/10.1002/(SICI)1099-0488(199808)36:11<1935::AID-POLB15>3.0.CO;2-8.

Low HY and Ishida H. 1999. "Structural effects of phenols on the thermal and thermo-oxidative degradation of polybenzoxazines." *Polymer.* 40, 4365–4376. https://doi.org/10.1016/S0032-3861(98)00656-9.

Macko, J and Ishida, H 2000, "Behavior of a bisphenol-A-based polybenzoxazine exposed to ultraviolet radiation." *Journal of Polymer Science Part B: Polymer Physics*, 38: 2687-2701.

Meador, MA 1998, "Recent advances in the development of processable high-temperature polymers." *Annual Review of Materials Science*, 28: 599-630.

Montserrat S, Andreu G, Cortes P, Calventus Y, Colomer P, Hutchinson JM and Malek J. 1996. "Addition of a reactive diluent to a catalyzed epoxy-anhydride system. I. Influence on the cure kinetics." *Journal of Applied Polymer Science* 61: 1663–1974. https://doi.org/10.1002/(SICI)1097-4628(19960906)61:10<1663::AID-APP6>3.0.CO;2-E.

Ozawa TA. 1965. "A new method of analyzing thermogravimetric data." *Bulletin of the Chemical Society of Japan.* 38: 1881-1886. https://doi.org/10.1246/bcsj.38.1881.

Pielichowski K, Czub P and Pielichowski J. 2000. "The kinetics of cure of epoxides and related sulphur compounds studied by dynamic DSC." *Polymer* 41, 4381–4388. https://doi.org/10.1016/S0032-3861(99)00694-1.

Pitchaimari G and Vijayakumar CT. 2014. "Studies on thermal degradation kinetics of thermal and UV cured N-(4-hydroxy phenyl) maleimide derivatives." *Thermochimica Acta* 575: 70– 80. https://doi.org/ 10.1016/ j.tca.2013.10.020.

Reiss G, Schwob JM, Guth G, Roche M and Lande B. 1985. in: B. M. Culbertson, J.E. McGrath (Eds.), *Advances in Polymer Synthesis*, Plenum, New York, pp. 27–49.

Schreiber. H, 1973, German Offen, 2,255,504.

Shamim Rishwana S, Mahendran A and Vijayakumar CT 2015. "Studies on structurally different benzoxazines: curing characteristics and thermal degradation aspects." *High Performance Polymers*, 27: 802-812 DOI: 10.1177/0954008314561806.

Shamim Rishwana, S and Vijayakumar, CT 2015. "Studies on structurally different benzoxazines: kinetics of thermal curing." *Roots International journal of multidisciplinary researches*, 1: pp. 99-102.

Shamim Rishwana, S, Mahendran, A and Vijayakumar, CT. 2015. "Studies on structurally different benzoxazines based on diphenols and diamines: kinetics of thermal degradation and TG-FTIR studies." *Thermochimca Acta*, 618: 74–87. http://dx.doi.org/10.1016/j.tca. 2015.09.006.

Shamim Rishwana, S, Pitchaimari, G and Vijayakumar, CT. 2016. "Studies on structurally different diamines and bisphenol benzoxazines: Synthesis and curing kinetics." *High Performance Polymers*, 28: 466-478. DOI: 10.1177/0954008315587125.

Toop DJ. 1971. "Theory of life testing and use of thermogravimetric analysis to predict the thermal life of wire enamels." *IEEE Transactions on Electrical Insulation*. 6: 2–14. 10.1109/TEI. 1971.299128.

Turpin, ET and Thrane, DT 1988. "Self-curable benzoxazine functional cathodic electrocoat resins and process." U.S. Patent. 4,719,253.

Vijayakumar, CT, Shamim Rishwana, S, Surender, R, David Mathan, N, Vinayagamoorthi, S and Alam, S 2013. "Structurally diverse benzoxazines: synthesis, polymerization, and thermal stability." *Designed Monomers and Polymers*, 17: 47-57. DOI: 10.1080/ 15685551.2013.797216.

Vyazovkin S and Dollimore D. 1996. "Linear and Nonlinear procedures in isoconversional computations of the activation energy of nonisothermal reactions in solids." *Journal of Chemical Information and Computer Sciences*. 36: 42-45. https://doi.org/10.1021/ci950062m.

Vyazovkin S and Sbirrazzuoli N. 1996. "Mechanism and kinetics of epoxyamine cure studied by differential scanning calorimetry." *Macromolecules*. 29: 1867–1873.https://doi.org/10.1021/ma95116 2w.

Vyazovkin S and Wight C. 1998. "Isothermal and non-isothermal kinetics of thermally stimulated reactions of solids." *A International Reviews in Physical Chemistry*. 17: 407-433. https://doi.org/10.1080/01442359 8230108.

Vyazovkin S and Wight CA. 2000. "Estimating realistic confidence intervals for the activation energy determined from thermoanalytical measurements." *Analytical Chemistry*. 72: 3171-3175. http://doi.org/ 10.1021/ ac000210u.

Vyazovkin S. 1997. "Advanced isoconversional method". *Journal of Thermal Anaysis*. 49: 1493-1499.https://doi.org/10.1007/ BF01983708.

Wang YX and Ishida H. 2000. "Synthesis and properties of new thermoplastic polymers from substituted 3,4- dihydro-2H-1,3-benzoxazines." *Macromolecules*. 33: 2839–2847. https://doi.org/ 10.1021/ma9909096.

APPENDIX: APPARENT ACTIVATION ENERGY

Flynn-Wall-Ozawa Method (FWO)

The FWO method is widely used for dynamic kinetic analysis and does not require any assumptions to be made about the conversion-dependence (Flynn 1996 & Ozawa 1965). The equation used for this method is

$$Ea = \frac{-R}{1.052} \frac{\Delta \ln \beta}{\Delta (1/T)} \qquad (A-1)$$

where R is the gas constant. In this method, plots of ln β versus 1/T give parallel lines for each reaction extent (α) value. The slope of these lines gives Ea, as per the following expression:

$$\text{Slope} = -0.4567(Ea/R) \tag{A-2}$$

Kissinger-Akahira-Sunose Method (KAS)

This method is based on the expression (Kissinger 1957)

$$\ln(\beta/T^2) = \ln(AR/Ea) - Ea/RT \tag{A-3}$$

Where β = heating rate, T = temperature, A = pre-exponential factor, R = gas constant, and Ea = activation energy under consideration. The plot of ln (β/T²) versus 1/T gives the slope, which equals Ea/RT by which the activation energy has been calculated.

Theory and application of model free kinetics approaches, starting from basic rate equation and ending in apparent activation energy prediction is discussed in the literature (Vyazovkin 1996 & Vyazovkin and Dollimore 1996).

Vyazovkin Method (VYZ)

The apparent activation energy can be determined at any particular degree of conversion by finding the value of Ea in equation 4

$$\sum_{i=1}^{n} \sum_{j \neq 1}^{n} \frac{I[Ea(\alpha),T]_i \beta_j}{I[Ea(\alpha),T]_j \beta_i} = 6 \text{ for } n = 3 \tag{A-4}$$

Friedman Method (FRD)

This is one of the differential methods (Friedman 1964) generally used to calculate Ea and equation is

$$\ln\left(\frac{d\alpha}{dT}\right) = \ln A + \ln f(\alpha) - \left(\frac{Ea}{RT}\right) \tag{A-5}$$

A linear regression graph is plotted between ln (dα/dT) and 1/T for different values of α from which Ea is obtained.

Corrected Flynn – Wall – Ozawa (C-FWO) Method

Farjas (Farjas 2011) introduced an iteration process that eliminates the necessity of the error table devised by Flynn. The accuracy of the conventional FWO method (Ozawa 1965) can be improved by adding iteration. Flynn improved the accuracy of the approximation by correcting the term -1.0518 in the slope of ln[p(x)] at $\overline{T}$, which is an average temperature over the transformation of the compound at different heating rates.

The equation that follows is,

$$\ln \beta_i - \ln \xi_{FWO} = \ln\left(\frac{AEa}{Rg(x)}\right) - 1.0518 \frac{Ea}{RT_i} - 5.330) \tag{A-6}$$

Now by taking the slope between $\ln \beta_i - \ln \xi_{FWO}$ and $-1/T_i$, the activation energy can be approximated by iterating the values. For the first iteration, the activation energy in equation 6 is assumed to the value obtained from conventional FWO Method. For the subsequent iterations, the value is taken as the activation energy obtained from the previous iterations.

Corrected Kissinger – Akahira – Sunose (C-KAS) Method

Just similar to the way how the approximation was corrected in conventional FWO method, Farjas (Farjas 2011) suggested the utilization of iterative procedure in conventional KAS method (Kissinger 1957) as well.

$$\ln \frac{\beta_i}{T_i^2} - \ln \xi_{KAS} = \ln \frac{AR}{g(\alpha)Ea} - \frac{Ea}{RT_i} \tag{A-7}$$

p(x) is taken from the Coats and Redfern approximation. xi represents the E/RT at $\bar{T}$, the average temperature over the transformation. Then this method too follows the same procedure as the Corrected FWO method to estimate the activation energy through iterative process.

Advanced Vyazovkin Method (A-VYZ)

For the small segment of mass degradation, at that specific reaction mechanism, the activation energy is going to be constant (Vyazovkin 1997).

$$J[Ea, T_\alpha] = \frac{A}{\beta_i} \int_{T_{\alpha-\Delta\alpha}}^{T_\alpha} e^{\frac{-Ea}{RT_\alpha}} dT \tag{A-8}$$

The rest of the procedure is followed as similar to the conventional Vyazovkin method and by minimizing the equation (Vyazovkin and Wight 2000) the activation energy can be calculated. The error calculation has been done using Advanced Vyazovkin.

Error Calculation in Advanced Vyazovkin Method

Vyazovkin (Vyazovkin and Wight 2000) has approached a statistical model to estimate the realistic confidence interval of the activation energy obtained through non-linear isoconversional methods. The method suggests that, for any heating rates at which the experiment is carried out, then the variance can be calculated using the following equation.

$$S^2(E_n) = \frac{1}{n(n-1)} \sum_{i=1}^{n} \sum_{j \neq i}^{n} \left(\left(\frac{J[E_a, T_i]}{J[E_a, T_j]} - 1 \right) \right)^2 \qquad (A-9)$$

The minimum variance is the minimum value for the activation energy. The confidence intervals can be identified for different confidence level preferred. This can give an approximate interval between which the activation energy occurs.

Parameter - ln A f(α)

From C-KAS method, the concurrent value of activation energy and the corresponding slope value are taken into consideration and intercept is obtained for the same from the equation 13. The intercept (INT) is taken for the left hand side of the equation and $-1/T_i$. The pre-exponential factor is calculated using the following relation (Liang 2015)

$$A f(\alpha) = \left(E_a / R \right) \exp(INT) \qquad (A-10)$$

Thermal Lifetime Prediction

The thermogravimetric curve can be used to predict the thermal lifetime of the material, provided the failure is a result of chemical reaction. Toop (Toop 1971) has proposed a rigorous mathematical solution to predict the thermal lifetime through the TG data. The activation energy estimated from the advanced method and the temperature integral of first order has been related to estimate the thermal lifetime.

$$\ln t_f = \frac{E_a}{RT\alpha} + \ln\left[\frac{E_a \, p(x)}{\beta R} \right] \qquad (A-11)$$

where t_f is the thermal lifetime of the materials, $p(x)$ is the temperature integral of first order, β is the heating program in °C min^{-1}, and $T\alpha$ is the temperature at the conversion α where the required set of properties of the material can be maintained.

In: A Comprehensive Guide to Formaldehyde ISBN: 978-1-53619-465-4
Editor: Natasja A. Bach © 2021 Nova Science Publishers, Inc.

Chapter 2

FORMALDEHYDE: POLYMER SURFACE CHEMISTRY AND DETECTION

Sanjiv Sonkaria and Hyun-Joong Kim[*]

Lab. of Adhesion & Bio-Composites,
Program in Environmental Materials Science,
Department of Agriculture, Forestry and Bioresources,
College of Agriculture and Life Sciences,
Seoul National University, Seoul, Republic of Korea

ABSTRACT

The global market trend predicts that the growth trajectory of the chemical "formaldehyde" will substantially increase production marking an unprecedented global economic expansion. The chemical profile of formaldehyde particularly in view of its reactive and volatile nature raises much concern about the potential chemical effects on disease progression due to inhalation at multi-levels in both natural and biological environments at bulk and nanoscale dimensions. As the global

---

[*] Corresponding Author's E-mail: hjokim@snu.ac.kr.

dependency on the use formaldehyde surges with no signs of slowing, much effort has been directed to investigate health-related issues and how nanotechnological advancement might play a greater role in in reducing the harmful impact of the chemical. The issues addressed in this chapter will converge upon a discussion and review of environmental and biological implications of the industrial growth of formaldehyde and a comprehensive evaluation of measures addressing the potentialities through the development of sensitive detection technologies.

Keywords: formaldehyde, nanomaterials, sensing technologies

INTRODUCTION

Formaldehyde formally known as methanal is the simplest hydrocarbon of the aldehyde group of chemicals and is considered to be ubiquitous in physical and chemical environments. It likely functions as a chemical precursor for many compounds produced in in living species from simple to more complex forms of life. Hence, its reputation as a primary source of carbon emphasizes its importance as a building block precursor in metabolic processes of life [1] and pathway regulation [2] with a possible evolutionary role [3, 4] and in synthetic intermediate to many important functional structures particularly at the nanoscale. Synthetically, the industrial relevance of formaldehyde has skyrocketed globally trending with economic growth on a global scale. This originates from its primary role as a fixative enabling biological materials to retain their 'life-like' state in the form of formalin or chemically hydrated formaldehyde. The global rise of formaldehyde correlates well with key adhesive characteristics in terms of binding strength, accessibility and durability and also its affordability as a polymeric material. Figure 1 demonstrates the applicability of bio-based formaldehyde complex resins in the bonding of wood for example where the adhesive performance is judged by mechanical strength, toughness and tensile strength aligned to greener properties. More importantly its role as a general preservative in diverse industries has been important from an economic point of view.

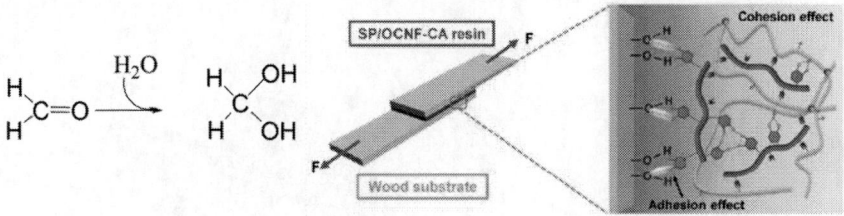

Figure 1. The application of hydrated formaldehyde (bio-resin) based materials for adhesion of wood. Adapted with permission from [5].

Formaldehyde as mentioned is notably be used for its adhesives properties for binding wood based materials and plastics and therefore is extensively utilized across diverse material surfaces to form chemically adhered layers in many industrial products. Some wood based furnishing for example comprise of hard polymer resins for manufacturing assembly parts for larger house-hold structures. The useful binding characteristics have also been used for textiles, domestic hygiene and medical products such as anti-bacterial materials, stationary items and more broadly as preservatives for many types of constituents. More so, formaldehyde remains a prevalent gas both indoor and out-door as it forms an important constituent of many house-hold items in wooden floors and furniture composed of different wood-types, interior and exterior wall coatings, thermal insulation materials, everyday textiles, in cleaning agents and chemicals as illustrated in Figure 2. Consumer products in the form of oral medicines have also been found to contain aldehyde derivatives. The nature of the pollutant and its wide spread use imply that it is not just restricted to the home but also the work place since the material is built into the fabric of numerous products.

High levels found in 'outside' environments are associated with car exhaust emissions and fuel sources with high formaldehyde content or through secondary events such as combustion of parent hydrocarbons and smoking. Its use in the agricultural and food industry has been particularly beneficial in food preservation.

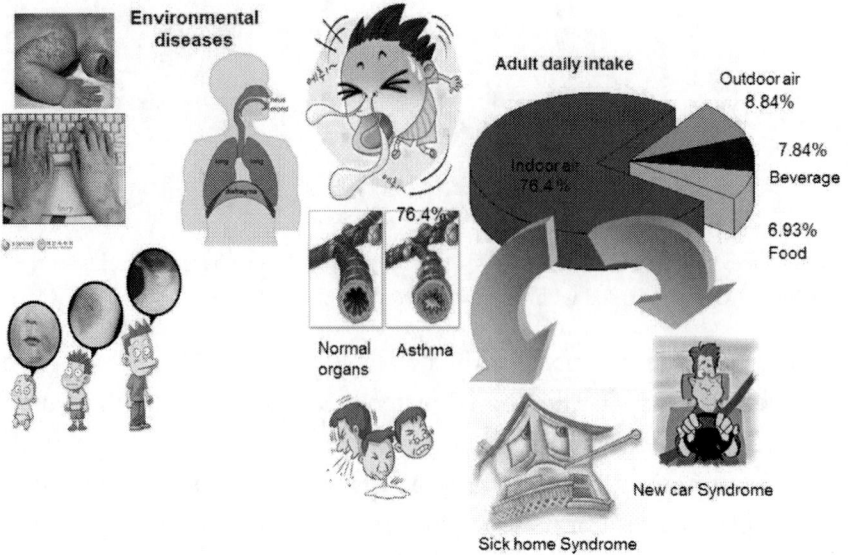

Figure 2. Environmental diseases from formaldehyde emissions in daily life.

A single dominating concern related to formaldehyde and its wide spread use among a variety of other chemicals is linked to its volatile and contaminating nature to the surrounding atmosphere. Formaldehyde exits in the gaseous state at room temperature but can polymerize to the solid state in the form of white powder [6]. The nature of the gas is complicated by the inter-conversion between the gaseous and solid states of matter questioning the stability of polymerized states in response to temperature changes in finished products. It can be readily released from its polymer paraformaldehyde form but other sources include dichlorophene, biocide and bronpol [7]. A long-term problem in the use of formaldehyde and polymer materials composed of formaldehyde is the persistence of vapors long-after the product manufacturing stage resulting in the slow release from buildings, furniture and numerous other products of over prolonged time scales into the surrounding air during the evaporation phase. Emission rates from fibre boards have also demonstrated to be highly correlative with temperature [8]. This has certainly been the case for long-term in door emissions of formaldehyde and its correlation with both temperature and humidity [9]. Air quality measurements quantifiably assert the existence of elevated levels of formaldehyde within in-door confined environments when compared to open environments. Formaldehyde is therefore considered as an air pollutant as most hazardous effects are mediated through the air. Abnormally high detection suggest that accumulative amounts of the gas amass from multi-sources which is consistent with indiscriminate use of the polymerized gas in day-to-day and longer-term manufactured products. The variance in indoor-to-outdoor readings is substantial with the prospect of a rising order of differences exacerbated by the contamination of gas emissions in closed environments such air cabins [11] as profiled in Figure 3. This increases the possibility of inhalation from emitted or absorption through the skin in the liquefied state.

If formaldehyde enters the body, it has the potential to undergo dissociation to formate or carbon dioxide which can be expelled from the body as liquid or gaseous waste products. However, body tissues are dose sensitive and formaldehyde can assert hazardous effects above threshold values which may vary from person-to-person. It is therefore classified as a public health hazard and toxicological evidence has accumulated over a number of decades. Excessive exposure to formaldehyde has strong toxicological consequences. Its highly reactive functional nature is associated with a number of pathological conditions and disease states. The lack of well-ventilated environments results in exposure to formaldehyde by inhalation of gaseous vapors, absorption of the hydrated form through the skin or consumption of contaminated food products if the contact time is adequate (Figure 4). The broad nature of diverse health effects from exposure to formaldehyde in humans [12] through these routes have shown populations to be at risk to immunological [13-16], neurological [17-22], oxidative stress [23-28], carcinogenesis [29-32], genotoxic [33-40], reproductive tissues and organs in both genders [41-47] and respiratory [48] complications.

Investigations on urine patient samples have revealed the ability of formaldehyde to alter small molecules establishing links to metabolic changes inside the body [49] and are involved in protein modification [50-54], stress induced mechanisms [55] and mimicking tau protein like aggregation [56].

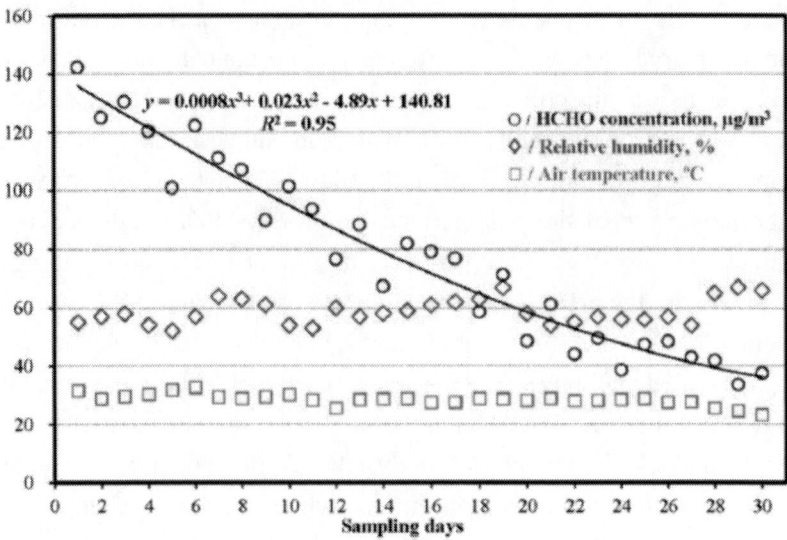

Figure 3. Data profiling of formaldehyde humidity and temperature variations in air cabins. Reproduced with from [10].

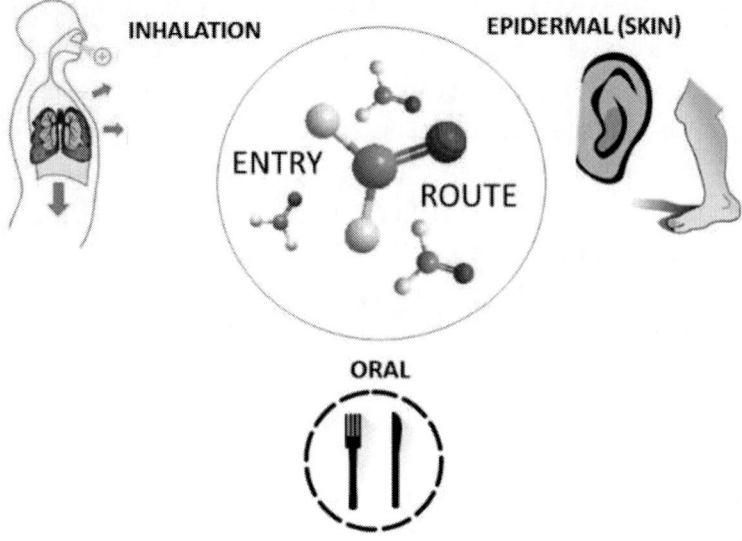

Figure 4. Different routes to the exposure of formaldehyde leading to the harmful effects inside the body.

Formaldehyde 53

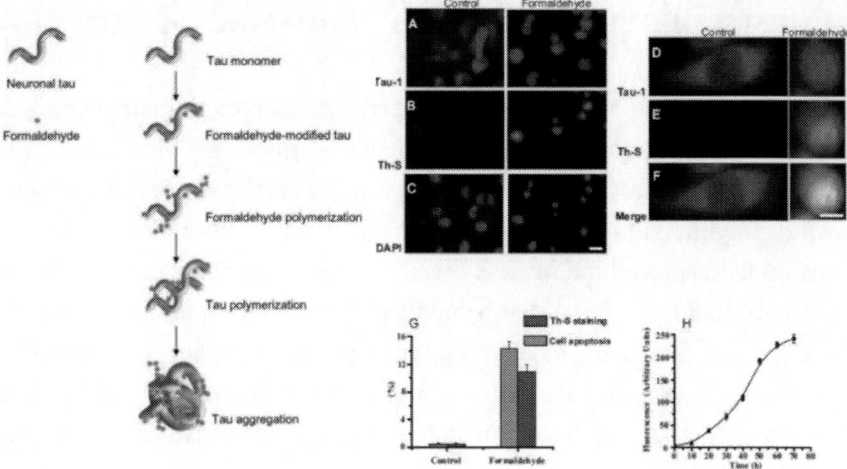

Figure 5. (Top panel left) The process of Tau protein aggregated assembly in the presence of formaldehyde. (Top panel right) Tau aggregation in the presence of formaldehyde in cells visualized by red fluorescence with Tau- antibody via ThS (amyloid-like aggregates). Blue fluorescence imaging are associated with nuclei of diminished size apoptotic arrest. (Bottom panel) Correlation between cell activity and antibody targeting signified by increased signals during apoptotic events relative to the control sample as a function of time Reproduced from [56].

Extensive investigations on the effects of formaldehyde have made clear that both domestic and occupational environments effect human health and defense barriers in the skin, respiratory tract and organs and tissues of the stomach are inadequate against the potent and reactant nature of the gas. It is freely absorbed into the respiratory tract and more limitedly by the skin but can enter through skin pores in the hydrated state. Food intake is another common way of entry into the blood which can be readily distributed among the organs, cells and tissues. Hence the severity of effects may be manifested by the period of exposure across short, intermediate and longer term durations. A more accurate correlation of the systematic effects on disease progression and mortality due to formaldehyde exposure could be ascertained by measuring and detecting formaldehyde levels in the surrounding atmosphere.

FORMALDEHYDE STABILITY IN PARAFORMALDEHYDE

Historically, establishing the polymerization rate of formaldehyde was highlighted as a challenging task on selected surfaces [57] in decades old investigations. Generally speaking however, the stability properties of gas-to-solid polymerization events of chemical components in formaldehyde for example or indeed the reverse case becomes an intricate problem in the basic understanding of material behaviour. This signals the need to explore characteristics like mechanical and thermal properties, kinetics and scale of structures (e.g., micro/nano) during polymer packing whilst undergoing phase transition. As discussed by Papon et al. [58], phase transitions are ubiquitous and phenomenological in nature but not very well understood conceptually. Formaldehyde aligned along material interfaces may result in property changes since polymerization likely occurs on surfaces and it is the interfacial complexities that may affect the ease of an ongoing phase transition in the material in response to environmental changes. An important concern in curbing formaldehyde as a health hazard is the manufacturing process of the product which relates to stability during the its polymerization phase. Commercial methods developed for the polymerization of formaldehyde have often used aqueous starting materials under supersaturating conditions of aldehyde [59]. However, adaptation of the method in the gas phase has been the methodological choice to aid the optimization of the polymerization process in terms of efficiency [60]. In wood fixtures for example, the surface stability of polymerized formaldehyde becomes a critically important question in reducing emissions from furniture. Formaldehyde monomers in the polymerized state requires that paraformaldehyde chains once formed should be not be free but bonded at the terminal regions. To achieve stability in the bonded state, surface adsorption is required. McGill and Söhnel [61] reported the findings of a computational study that modelled the binding of paraformaldehyde to TiO_2. Here, formaldehyde was shown to participate in both weak coordinate and stronger dimerization bonding to the TiO2 surface. Figure 6 shows the dimerization bonding of formaldehyde via Ti and O atoms caters for a stronger bonding environment which may implications in reducing environmental pollution and slow poisoning by inhalation.

ADSORPTION OF FORMALDEHYDE

Control of formaldehyde pollutants through adsorption from the surrounding air by adsorbent materials particularly indoors from trapped air has been considered as a possible effective route remove the harmful effects on the human body. This is becoming an important directive as many house-hold items emit the gas from electronic equipment, furniture, wall paints, floor polish and cleansing chemicals routinely used in confined environments. If ventilation is poor in rooms with little no measures in place to remove toxicity build of volatiles in the air we breathe, the day-to-day accumulative effects of gas inhalation can easily escalate from levels of low toxicity to chronic elevated concentrations. This often leads to the state of 'sick building syndrome' that manifests symptoms of fatigue, dullness and low libido [62].

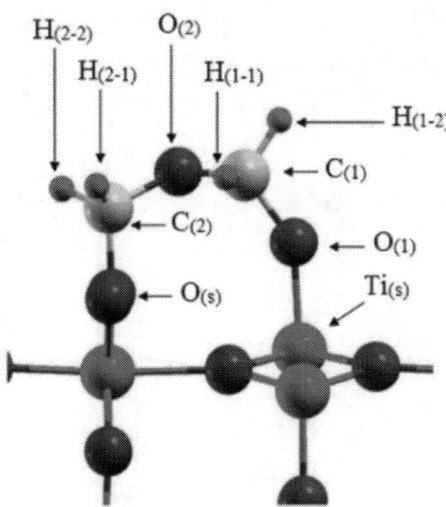

Figure 6. Computational model of the favourable binding of formaldehyde to TiO_2 surface in the gaseous adsorbed phase [61].

The application of adsorbent materials have recently been reported. Microporosity intrinsic to the nature of indoor materials has been suggested and investigated as a viable solution to limit the release of toxic gaseous vapors in the home or office environment. Such innovative materials have been evaluated by sorptive building materials test (SBMT) and researchers have found celite as a very effective adsorbent reducing HCHO allowing reduced concentrations of the volatiles to 0.2 ppm under a loading factor of 0.4 m^2/m^3 [63]. Figure 7 shows the application and evaluation of HCHO adsorption and its retainment by selected materials in this study. Similarly, advanced functional materials have been designed in the form of metallorganic frameworks to perform under ambient conditions. One such example is UiO-66-NH2 which exhibited an adsorption capacity of formaldehyde of 69.7 mg g^{-1} and was mainly attributed to regions between framework linkers and the affinity of hydrocarbon chains at the vicinity of metal sites and their association with carbonyl groups [64].

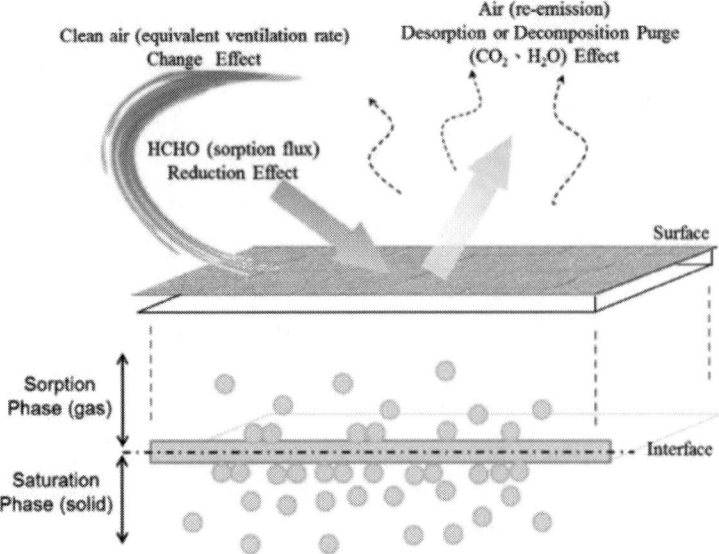

Figure 7. Material evaluation for the effective adsorption and retainment of the contaminant formaldehyde released into the local environment from household objects. Reproduced from [63].

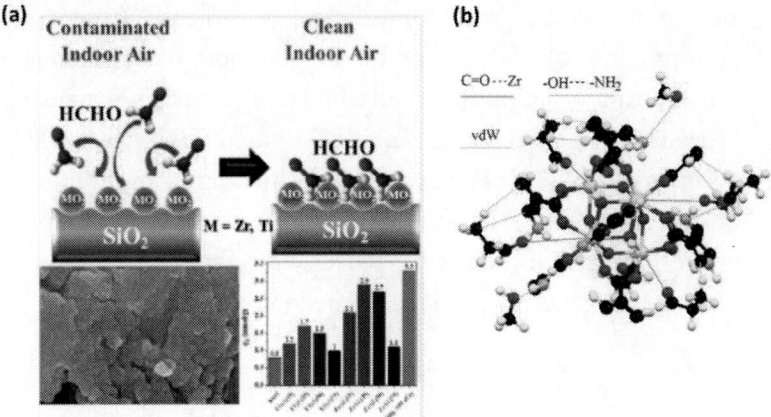

Figure 8. (a) An example of removal of HCHO vapors using ZrO_2/SiO_2 and TiO_2/SiO_2 metal-organic frameworks (MOFs) (b) Molecularly adsorbed contaminates of Zr clusters permitting their active removal. Reproduced with permission from [69] and [64].

Further investigations in this direction have provided support a range of materials including mesoporous silica organic frameworks (Si-(OH)-Al), zeolites and magnesium and aluminum complexes for formaldehyde capture. A copper complexed material designated HKUST-1 provide a promising example of porous materials to accommodate and retain HCHO molecules. The observation however, that humidity is a major player in reducing gas adsorption capacity suggests that the dimensional pores in these porous materials are very small scale to be effective in the entrapment of formaldehyde [65].

Porosity as an essential feature of all applicable materials include porous carbon and silica materials [66], activated carbon [67], diatomite, bentonite and zeolite [68], mixed metal oxides ZrO_2/SiO_2 [69] and TiO_2/SiO_2, covalent-organic polymers [70] and amine-functionalized mesoporous silica materials [71] among others. Further, the removal of formaldehyde by photocatalytic oxidation was observed to occur proportionally to the semiconductor content and shown to be dependent on surface contact [72]. Correlations between functional group chemistry at the nanoscale and nanopore size in relation to the physical, chemical and functional nature of formaldehyde is proving to be an important cornerstone to the development of efficient adsorbents. Computational analysis can provide valuable insight in mapping both these parameters to chemical composition tuning the applicability of materials for adsorptive uptake of HCHO in preference to water [73] based upon guest-guest and guest-host interactions [74]. For example, positioning of pyrrolic nitrogen on carbonized resins could be adapted to heighten the interplay between electrostatically polarized groups in HCHO with nitrogen suggesting that high surface area alone can have a heightened impact [75].

DETECTION AND QUANTIFICATION OF FORMALDEHYDE

In view of the considerable public health issues in the use of formaldehyde based on toxicological and epidemiological studies, scientists, health officials and governmental regulatory bodies and toxicologists have made concerted efforts and recommendations to limit the use of the substance and proposed safety limits of formaldehyde concentrations exceeding tolerable levels. This is a rising concern with the scarcity and lack of new epidemiological data [76], early warning detection systems should accompany the growth trajectory of formaldehyde products which are exponentially rising with product demands. The detection of levels of formaldehyde exceeding recommended safety levels require detection systems that can sample air quality and free formaldehyde in the air, water and land. Enforcing legal requirements that defines short and long-term limits of exposure in domestic and occupational environments has placed economic pressure on the chemical industry. Since formaldehyde was formally categorized as a health hazard and pollutant in the early sixties, serious efforts have directed researchers to consider increasingly innovative ways to use materials for the quantitative detection. In line with the World Health Organization (WHO) defining indoor HCHO levels not to exceed 0.1 mg m^{-3} (80 ppb) [77], innovation is facing a considerable technological task to enable safer habitats and living environments while the global utilization of formaldehyde is witnessing a surge in production. Additionally, the molecular scale of detection is introducing a new dimension (Figure 9) to these challenges compelling a new wave of materials engineering at the nano and quantum scales to uncover structure/functional relationships to address these concerns. This requires the development of complex functional materials. Here, we present a selection of detection technologies from early on to the present time.

EVOLUTION IN THE DETECTION FORMALDEHYDE

As early as 1945, spectrophotometry was suggested as a method for the quantitative detection and measurement of formaldehyde [79]. This was based on a colorimetric assay involving a reaction of formaldehyde with chromotropic acid requiring the discharge of a purple colored product. The accuracy of demanded optimal working conditions to develop the colorimetric change with high purity chromotropic acid to minimize errors. The method however is at best semi-quantitative requiring a large window of time for the color development against the availability of standards. The method itself is based on microgram quantities and would not be suitable for low emission detection in the range of ppm. Concerns over air contamination of formaldehyde and chromotropic acid was recognized in 1952 along with 2,7 dihydronapthalene which still remained the main reagents of focus in spectroscopic investigations due to their rivaled readability in range of 0.05 – 2.0 µg/ml obeying linearity of the beer-lambert law [80]. A decade later, cationic and dicataionic dyes were suggested as alternatives to chromotropic acid increasing the order of sensitivity by 2.5 times [81] and both required the 'release' of a colorimetric product to indirectly quantify HCHO molecules. In the modern era, spectroscopic sensitivity is subject to dilution factors and molar absorptivity which obey the beer-lambert law. Table 1 compares the spectroscopic methods employed for the detection of formaldehyde in the air [82]. Each of these methods entail both drawbacks and benefits to their use but the ease, relatively inexpensive, accessibility and selective nature of spectroscopic approaches allow them to be widely used and preferred both in technicality and usability. Gasparini et al. [83] addressed the greener aspects of the use of chromotropic acid by using magnesium for the complexation generating a spectroscopically detectable intermediate Mg^{2+}/cycl-o-tetra-chromotropylene with HCHO molecules in the range of 3 to 11 mg L^{-1}. This enabled a more ecologically acceptable spectroscopic reagent contributing to a greener environment (Fluoral-P) as a colorless agent which readily reacts with formaldehyde in brain tissues of rats to generate a spectrophotometric measurement of 3,5-diacetyl-1,4-dihydrolutidine (DDL) with limits of detection of 0.5 μM and 2.5 μM [84]. Spectroscopic methods have found utility in standardizing the development of non-spectroscopic approaches to the quantification of HCHO (Figure 10) capable of detection. For example, Lamarca et al. [85] used an instrument for the applicability of spectroscopic procedure for biological tissue analysis has advanced the technique for utility potentially for clinical samples. For example, the availability of 4-amino-3-penten-2-one detection of HCHO in cosmetics was based on gas-diffusion microextraction integrated with a smartphone reader. Validation of the new method by spectroscopic analysis as the reference method is shown in Figure 11. The smart phone method was able to generate signals with the same degree of sensitivity to the reference method in range that was quantifiably agreeable (range limit ~ 0.200 mg kg^{-1} and 0.500 mg kg^{-1}).

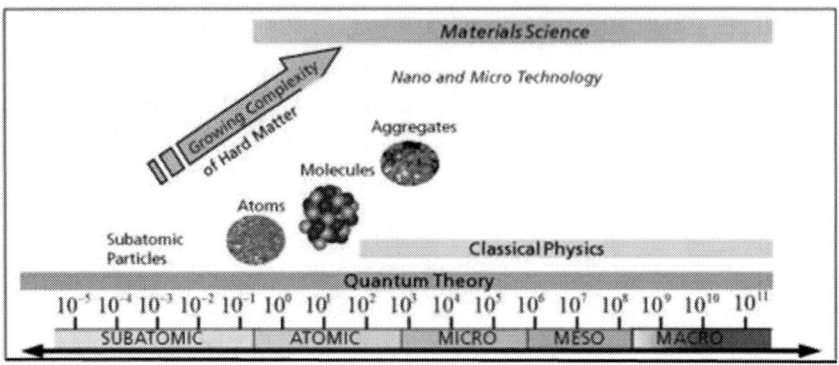

Figure 9. (a) Hierarchical scale of particles in materials engineering. Reproduced from [78].

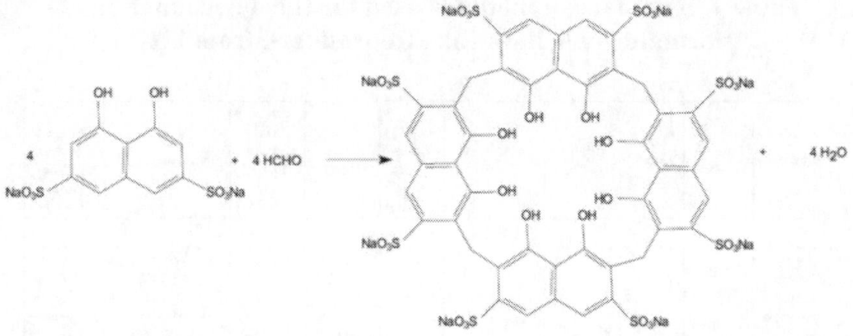

Formaldehyde

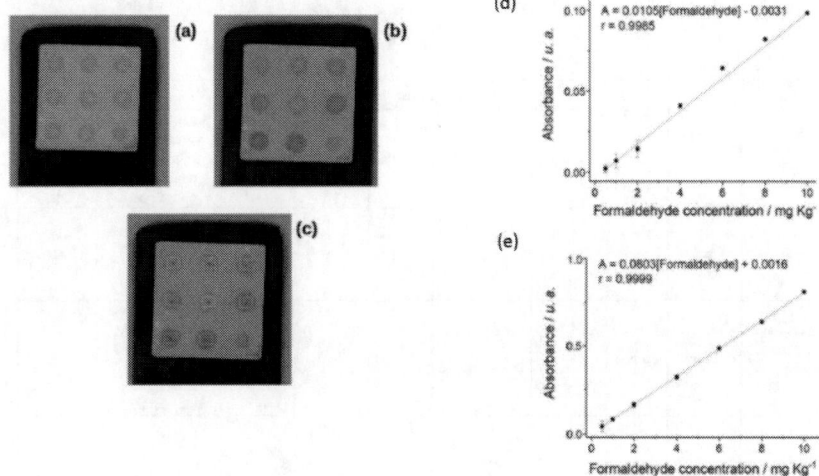

Figure 10. Modification of chromotropic acid to Mg^{2+}/cyclo-tetra-chromotropylene. Reproduced from [83].

Figure 11. (a-c) A direct route for the quantification of formaldehyde in cosmetic products by a smart phone approach and validation of detection (d-e) spectroscopic determination. Reproduced with permission from [85].

Table 1. Summary of methods used for the determination of formaldehyde in the air. Reproduced from [82]

Method	Sensitivity	Time requirements	Colour stability	Interferences	Wavelength	Conc. Limit	Financial requirements for chemical	Disadvantages	The most common application
Phenolic type compounds									
Chromotropic acid with H_2SO_4 (CA, Original) [13,15]	1.57×10^4 $1.mol^{-1} cm^{-1}$	Long time (1 h)	24 hod	SO_2, NO_2, NO_3, hydrocarbons	580 nm	0.19-0.5 ppm	Low cost	Use of potentially hazardous, corrosive sulphuric acid, interference of metal ions, sugars, other, long waiting time	The atmosphere, combustion effluent
Chromotropic acid with HCl and H_2O_2 [19]	$(1-71+/-0,02) \times 10^4$ $mol.l^{-1} cm^{-1}$	Long time (1 h)	-	-	575 nm	-	Low cost	Long waiting time	Atmosphere, combustion effluent
Schiff reagens									
Fuschin [56]	3.5×10^3 $L.mol^{-1} cm^{-1}$	Long time (2 h)	30 min	-	550 nm	0.83 ppm	Low cost	Beers low is not satisfactorily followed and require long waiting time	Air
Fuschin with acetone [3,4]	3.5×10^3 $1.mol^{-1} cm^{-1}$	Long time (3 h)	30 min	SO_2, NO_2, alcohol, phenol	560 nm	5 ppb	Low cost	Long waitting time	Air
Pararosaniline original [21]	Low	15 min	-	Acetaldehyde, propionaldehyde	560 nm	0.01 $\mu g.ml^{-1}$	Expensive	Tetrachloromercu-rate is costly and toxic, low sensitivity, interferences from aldehydes	Air
Pararosaniline mercury free [28]	1.88×10^4 $1.mol^{-1} cm^{-1}$	60 min	120 min	Low molecular weight aldehydes, sodium sulfite, potassium cyanide, sodium nitrite, hydrogen peroxide, hydroxylamine	570 nm	25 ppb/60 l	Low cost	Long waiting time, interferences	Air from the nonindustrial indoor environment, water
P-aminobenzene (original) [3]	4.5×10^4 $1.mol^{-1} cm^{-1}$	20 min	90 min	Organic and inorganic compounds	505 nm	0.08-0.48 ppm	Low cost	Interferences from co-pollutants	Air

Method	Sensitivity	Time requirements	Colour stability	Interferences	Wavelength (nm)	Conc. Limit	Financial requirements for chemical	Disadvantages	The most common application
P-aminobenzene sulphonic acid [30]	1.5×10^3 $l.mol^{-1} cm^{-3}$	20-25 min	60 min	Organic and inorganic compounds	510 nm	0.072-0.38 ppm	Low cost	Interferences from co-pollutants	Air, automobile exhaust
Hydrazones									
MBTH [3,12,37,39,]	6.5×10^4 $l.mol^{-1} cm^{-3}$	10 min	4 h	Other lower aliphatic aldehydes	670nm	0.05 ppm	Very expensive	Stability is not good as in acetylacetone method, interferences from lower aliphatic aldehydes, expensiveness	Epoxypropane, seawater, outdoor and indoor air
DNPH [58,60]	1.76×10^4 $l.mol^{-1} cm^{-3}$	Long time	-	Ozone oxides, NO, NO_2	357 nm	0.012-2.00 and 18.00-372 $g\,ml^{-1}$	Expensive	Need to use HPLC before UV/VIS spectroscopy, expensiveness	Polysaccharides, Automobile exhaust water, air pollution
Hantzsch reaction									
Acetylacetone (AA) [59]	High sensitivity	10 min + cooling	-	NO_2, SO_3 ions; acrolein, acetaldehyde at 100pm,	-	0.1 ppm	-	Slow procedure, acetyl acetone is more expensive and contaminates environment	Air, fabric owning
Other reagens									
AHMT [57]	2.1x greater than AA method	20 min	-	Acrolein, acetaldehyde, propionaldehyde, n-butyraldehyde, at 1ppm	550 nm	0.05 ppm	Very expensive	Expensiveness, interferences from other aldehydes	Air, water
Phloroglucinol in alkaline conditions [55]	2.1×10^4 $L.mol^{-1} cm^{-3}$	Low	Unstable	Free from interferences of aldehydes	470 nm	0.1-2 $\mu g.ml^{-1}$	Low cost	The colour is unstable	Air, distillery waste, polluted river water, coke oven effluent

APPLICATION OF NANOPARTICLES AS FORMALDEHYDE SENSORS

Nanoparticles exhibit an array of properties that are unique to their dimensional size and atomic configurations. Such properties open up new opportunities to screen and profile new recognition features and atomic efficiencies for catalysis for example. Such intrinsic properties are often emergent from the transition from bulk to nano- or sub-nano scales. Scale reduction bring new inherent reactivities that are intrinsic to physical and chemical processes used in the assembly process. Interatomic spaces that arise from geometric constructions at the smallest scales and the accompaniment of charge distributions can play a pivotal role in the capture of diffusible small molecules with capabilities to entrap them. The structures can exist at very complex interfaces complicating our understanding of structure functional relationships. In materials of low dimension, parameters such as material shape, size, morphology, composition and newly emergent properties that do not exist in the bulk phase become important facets to developing innovative technologies in gas sensing. Figure 12 depicts the free energy changes that accompany nucleation processing directing the self-assembly of morphologies and interchanges. Such morphological properties are a useful addition to gas sensing applications and manifest in different physical and chemical functionalities at the nano and quantum scales. We aim to provide some insight in the remarkable and special utility in their interaction with formaldehyde.

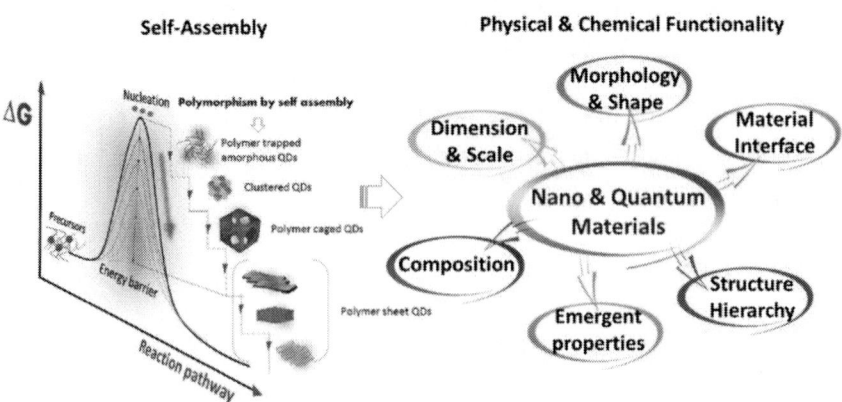

Figure 12. Pathways to evolving self-assembled structures and defining physical and chemical functionality of materials. Reproduced from Khare and Sonkaria [86].

Formaldehyde

One example demonstrates how nanoscale nano-gold particles divergent in shape and morphology behave intrinsically differently as sensors for HCHO the properties enabled by the setup shown in Figure 13. This was signified by the difference in the plasmonic properties for spherical particles and nanorods shaped morphologies. In this case, spherical particles were associated with an increase in sensing activity against formaldehyde from 25 to 35 nm indicating a strong size dependency for spherical shaped particles [87]. Shape induced changes in surface properties due to alteration in morphology revealed that nano-rods generated two absorption peaks at 578 and 650 nm mapped by transverse and longitudinal surface plasmon resonance (SPR). This response was distinctive to the spherical particles resulting in a decrease in the peak signals in the presence of decreasing concentrations of formaldehyde suggesting that new functionalities could be liberated from plasmonic properties by a change in surface morphology. This leaves the potential to further configure novel sensing strategies for portable sensing technologies.

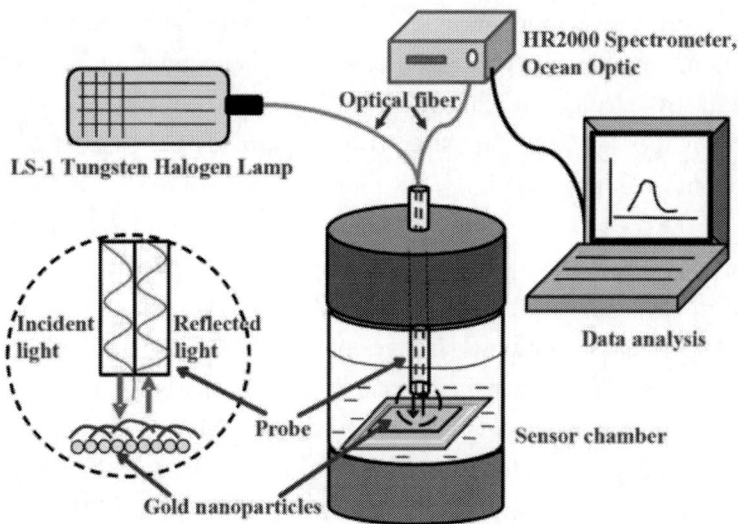

Figure 13. Setup used for formaldehyde using morphologically different gold nanoparticles. Reproduced from [87].

A more elaborate example of nano-engineering based on multi-functionality in the sensing applications of nanoparticles comes from the complex interface of several nanoparticle elements. True multi-functionality arises when each component is able to exhibit characteristics unhindered by the complexation of the other making possible a dual operative sensing device. This feature in nanomaterials engineering was informed by a study reported by Wang et al. [88] where Au@ZnO were shown to bind both formate and water molecules indiscriminately with similar binding energies. The inclusion of ZIF-8 to Au@ZnO favored better HCHO selectivity following full saturation of the ZnO surface with HCHO/H$_2$0 with the effect of nullifying the shallow signal associated with ZnO surface and thereby leading to the activation of the broader and highly responsive Au@ZnO bound ZIF-8 surface as judged by the Ra/Rg ratio (= sensor resistance in air/resistance during exposure to HCHO).

The lower limits of detection of formaldehyde by material improvements in new sensing systems are currently being addressed. In a very recent effort, researchers have used amorphous nanoparticles in which Eu atoms were deemed to be essential adsorption sites for $Eu_{0.9}Ni_{0.1}B_6$ [89]. Figure 14(a) shows nanoparticles around 6.5 nm and shows a linear correlation in the presence of increasing concentrations of HCHO measured by change in capacitance (ΔC) between the presence and absence of the gas. The sensitivity of Eu particles increased from their presence in EuB_6 to $Eu_{0.9}Ni_{0.1}B_6$ but found to be irresponsive to Ni, B, and B_2O_3 and the lack of $^2H_2$, ^2H_2O, or $^{13}CO_2$ emissions in $Eu_{0.9}Ni_{0.1}B_6$ but not EuB_6 showed the gas molecules better occupied the interparticle spaces of the adsorbent molecules. The sensitivity to HCHO intake is shown in Figure 14(b) demonstrating a stronger preference towards binding HCHO gas molecules and the sensing capability of the particles is profiled in Figure 14(c) in concentration dependent manner from 2 ppm –20 ppm by monitoring the gas input to output ratio. Further, the cooperative role of Ni in $Eu_{0.9}Ni_{0.1}B_6$ facilities the ease of electron transfer at the site of HCHO adsorption from Eu to formaldehyde establishing that dimensional order in space is an important factor during the interaction of $Eu_{0.9}Ni_{0.1}B_6$ with the gas molecules.

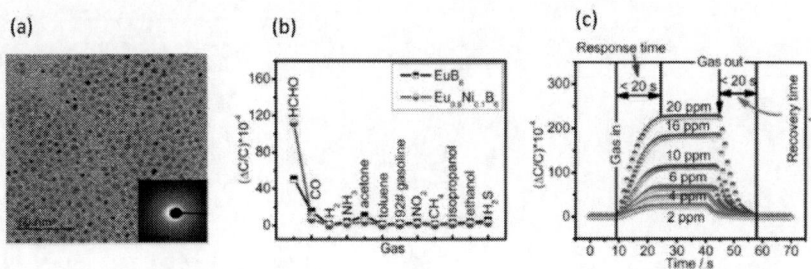

Figure 14. (a) HRTEM image of $Eu_{0.9}Ni_{0.1}B_6$ nanoparticles and the (b) gas selectivity of $Eu_{0.9}Ni_{0.1}B_6$ towards HCHO in comparison to EuB_6 particles (c) Detection of formaldehyde as a function of time versus the ratio between capacitance change and capacitance in the absence of the gas. Reproduced with permission from [90].

At the nanoscale, understanding the interactive chemistry of formaldehyde has become an integral part of developing new multi-complex molecular systems that exploit functionalities and dynamics of coordinate chemistry. More than a decade on, functionalised polymer films on pH paper have shown great utility in low sensitive detection of formaldehyde [91]. Despite the simplicity, remarkable sensitivity (20 – 250 ppm) and speed of detection with visible color change as a portable device, new chemistries have been explored. A particularly useful feature in advancing new strategies of greater sensitivity of vapor detection has been fluorophore excitation chemistry. For example, the synthetic labeling of pyrene molecules with a fluorophore (PPB) and its subsequent aqueous phase reaction with a sugar monosaccharide provided a platform for formaldehyde detection [92]. The equation in Figure 15 shows that the formation of an excimer proceeds from an excited pyrene monomer which combines with a ground state pyrene. Also, the fluorescence intensity profile in Figure 15 shows a lower energy (red-shift) peak-shift with increasing absorbance change at 490 nm which correlates well with an increase in HCHO accompanied by a decrease in pyrene emission. The dynamics is controlled by the re-arrangement of HCHO molecules on the PPB scaffold allowing pyrene molecules to be positioned closer to each other and thus altering the fluorescent properties of the complex.

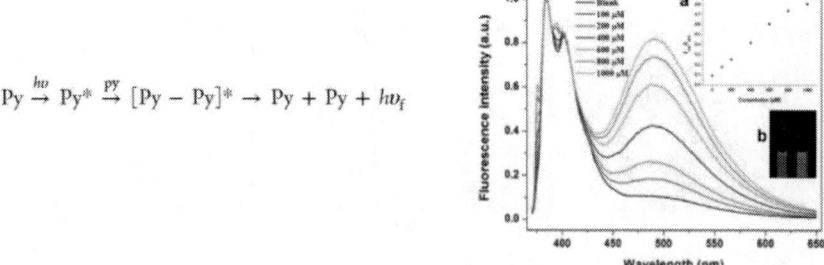

Figure 15. The LHS equation shows the mechanistic step for excimer formation and specificity in binding formaldehyde molecules resulting in a concentration-dependent increase in the fluorescence signal Reproduced from [92].

The strategy is one example of 'aggregated' and 'disaggregated' states and demonstrate how the dynamics of the complex permit interchangeability while being coupled to changes to the fluorescent states. To broaden the limits of detection from 1–100 ppm, a hydrophilic hydrazino-naphthalimide functionalized chitosan polymer was chosen as a specific HCHO detector through the activation of a naphthalimide fluorophore triggered chemically [93]. Tang et al. have overcome some of the challenges in using fluorescent probes to track the adsorption of formaldehyde inside biological tissues based on the design of 'one-photon excitation' probes. Here, the team exploited hydrazine-aldehyde chemistry that readily forms methylenehydrazine which is chemically durable and retains its hydrazine functionality when trapped inside a two-photon responsive dye namely, 1,8-naphthalimide [94]. The resulting Na-formaldehyde complex permitted electron transfer allowing the 'non-activated' fluorescent state of 1,8-naphthalimide to be 'switched on' by the interaction with formaldehyde shown by an instantaneous rapid peak evolution at 543 nm. Cells treated with formaldehyde were highly detectable demonstrating its non-toxic nature and considerably high specificity for formaldehyde as low as 1 μM. However, more elaborate molecular recognition strategies sensitive to formaldehyde continue to be reported making use of chromophore aggregation and photocatalysis using aggregation-induced emission (AIE) [94]. In this case, a change in luminescence that results from the aggregated character of photo-reactive structures was used for sensing properties using hydrazine chemistry (thioacetylhydrazine groups) for chemically coupling formaldehyde and $\pi \cdots \pi$ statcked bis-thioacetylhydrazine-functionalized pillar[5]arene groups to accommodate fluorogen in an AIE directed mechanism and the catalytic properties of (CF3SO3)2Bi for further improving sensitivity. Using this setup, the authors reported a sensing platform with a detection performance of formaldehyde to the lowest limits of detection of 3.27 nM within 7.5 s. For a range of different aldehydes tested for fluorescent activity the complex showed the highest structural correlation with formaldehyde (sample 14) shown in (Figure 16a) and the emission was observed to be associated with a stronger blue light effect than red (Tyndall effect) (Figure 16b, c). The specificity of binding towards formaldehyde is shown increasing anchoring of HCHO molecules and the associated fluorescence emission (Figure 16d).

Understanding key principles in molecular orientation between two interacting interfaces is gaining tremendous great importance in gas sensing technologies. For example, electrostatic, hydrophobic, hydrophilic, hydrogen bonding and $\pi - \pi$ stacking is critical to the aggregation and disaggregation interchangeability as discussed previously both in the same phase or multi-phase scenarios. Hence, accessibility to functional states and their implementation in such technologies requires an appreciation of molecular interaction of chemical groups at nanometer and micrometer thickness [95]. This is exemplified by the gas sensing properties of nickel oxide nanoparticles. The semiconductor properties of NiO is strongly coupled to conducting characteristics that result from chemisorption/ physisorption of adsorbed oxygen molecules that is directly seeded in temperature related effects. The enhancement in the molecular adsorption of oxygen is directly associated with the increase in electron traps and increased surface area of lower sized nanoparticles offering surface resistance to conducting electrons. This intrinsic behaviour of semiconductor chemistry of nanorose was pursued as a sensor based on resistance upon complexation with formaldehyde resistance and the resistance of the sensor measured in air (R_g/R_a) (Table 2). The sensing range achieved superiority over other reported NiO based sensors in a working range of 50-1000 ppb with unprecedented recovery times of 47.2 and 8.2 s [96]. In a separate study, the fabrication of NiO thin film morphologies of thickness 150 nm yielded a gas sensing capability measured by its resistance towards formaldehyde with an order of sensitivity between 5 to 20 ppm exceeding the performance of the same thin file morphology of 300 nm [97]. This clearly established the effect of grain size as a fundamental property in capturing vaporized HCHO molecules via diffusion in conjunction with other material characteristics such as porosity. The increased adsorption of HCHO within the charged spaces of NiO and the effect on the conductivity of the film surface is counterbalanced by the diminishing size of NiO particles permitting an increased surface area that results in broad linear sensing response as a function of resistance. Further, the fast response time of a laboratory synthesized based ZnO preparation and its associated grain property proved to a tunable parameter as an impedance based sensor for formaldehyde. The lower grain size of 26 nm marked the origin of a morphologically relevant sensor conducive to low level interacting concentrations of formaldehyde at a concentration range of 100 – 800 ppm [98]. Surface specificity resides in the congruency of 3-dimensional assemblies with inter-crystalline spaces and design features for better detection as shown for ZnO particles [99]. Further, nickel particles have been shown to undergo 'thin film' formation through coalescence which increased with increasing deposition time. Subsequent layering and cloistering of particles formed the basis of an electrochemical sensor for formaldehyde with a linear range of from 0 - 6.5 mM although a detection was reported to be in the micro molar range corresponding to less than 5s response time [100]. In the application of NiO materials as effective formaldehyde gas sensors, the application of different synthetic processes clearly establishes performance related differences in sensing abilities emphasizing that morphological characteristics if suitably tuned to the physical and chemical properties of the target contaminant, as small as 1ppm of HCHO is detectable [101].

Formaldehyde

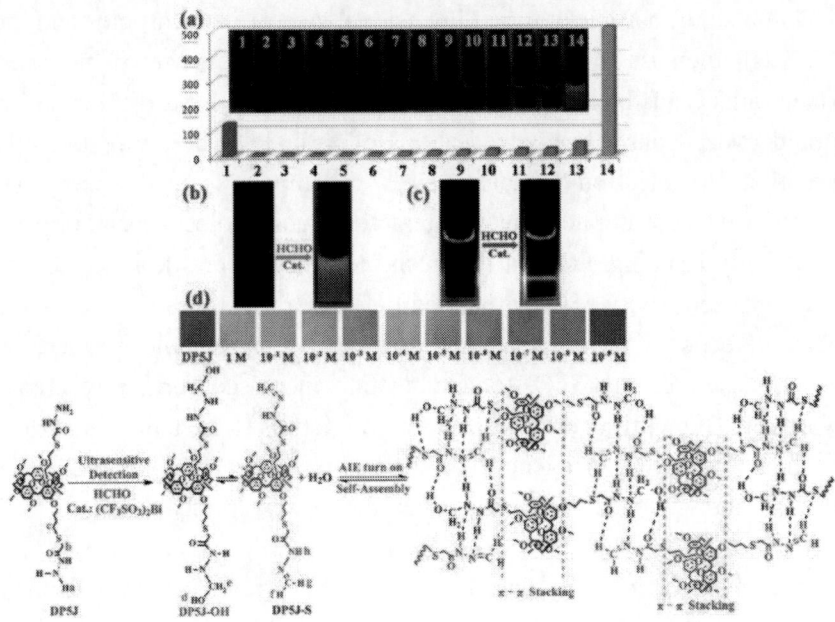

Figure 16. (a) Emitted fluorescence measured against a range of aldehydes (b) Fluorescence imaging in the presence of formaldehyde in DMF (c) Imaging of the Tyndall effect supported by stronger blue light emission under excitation (d) The effect of increasing formaldehyde concentration on fluorescence imaging. A structural representation of the molecular anchoring of HCHO molecules to the fluorescent scaffold. Reproduced with permission from [94].

Table 2. Sensing parameters based on the conductance resistance of NiO nanoparticles measured at 230°C. Reproduced from [96]

Concentration (ppb)	$R_g(k\Omega)$	$R_a(k\Omega)$	$S = R_g/R_a$	T_{re} (s)	$T_{re}(s)$
50	109.6	62.7	1.75	47.2	8.2
100	181.9	67.3	2.70	46.4	8.4
200	204.6	64.3	3.17	45.3	11.6
500	205.6	63.8	3.22	58.5	11.3
800	210.5	62.5	3.37	55.2	12.1
1000	211.9	61.8	3.43	53.7	13.3

Diminishing particles to the dimensional size of nanoparticles and the subsequent increase in surface reactivity as a consequence of increased surface area can better interface with the reactive surface such as formaldehyde. Charged polarized atoms of formaldehyde are particularly susceptible to electron deficient sites or unpaired bonds and their accessibility to unsaturated site of interacting guest molecules can provide high affinity bonding sites. In this respect, studies were designed to test surface gas selectivity of SnO_2 nanoparticles as a function of the available oxygen defect sites favoring surface adsorption of target molecules. Gao et al. [102] used 20 nm SnO_2 to demonstrate increased surface reactivity towards HCHO with superior gas sensing properties to the bulk counterpart reaching a capturing efficiency of 50 ppm. Structural surface properties was also shown to exhibit low concentration adsorbidity sensitive to 1 ppm.

Advancement in improving device design for reducing formaldehyde toxicity has witnessed significant steps in adopting bioinspired approaches both recognition and capture based nanotechnologies. Mimicking biological patterns in the assembly of biomaterials at the nanoscale has been challenging to defeat problems in selectivity and smarter properties [103]. Wu et al. [104] fabricated nanochannel architecture internalized by carboxylate groups providing a negatively charged interior using polyethylene terephthalate by a simple etching process of the polymer lining with ethylenediamine (EDTA) and carbodiimide/N-hydroxysulfosuccinimide sodium salt (EDC/NHSS) coupling strategy to activate available COOH sites. An 'open' state achieved using this method facilitated the entry of HCHO molecules and their subsequent immobilization via covalent NH_2^+----COO^- covalent bonding networks. This arrangement liberated a unidirectional flow of ionic currents resulting from the 'transmembrane-like' migration of HCHO molecules through nanometer sized pores imitating a closed state through the occupation of the channel interior pockets. The setup constituted a 'smartly-driven' biomimetic process for the specific detection and removal of formaldehyde.

CONCLUSION

Significant progress has been made in formularizing effective nanomaterials using various fabrication approaches towards the detection of formaldehyde. The demand for green materials that potentially reduce the inhalation of formaldehyde fumes [63] and other VOCs will play a critical role in this direction. However, the enormity in the scale of VOC sources of formaldehyde emissions [105] mandates better detection and risk management strategies to substantially manage and reduce health risks due to accumulated toxic inhalation. Although there is a growing realization that morphological patterns of materials at the nanoscale have a strong correlation with their performance as detectors, surveying the ability of material properties to performance has yet to be fully understood. Since the contaminated forms of formaldehyde can exist in both aqueous and gaseous states, it is imperative to reconcile these efforts in material design for optimal detection under different environmental conditions including air, water, food and biological samples. As our understanding of material properties expand aided by computational predictive tools, it is expected that the current knowledge will narrow and a more random study of investigative materials will be replaced by a more systemic approach to more efficiently reduce the harmful effects of formaldehyde and its complexes.

REFERENCES

[1] CHAPTER 1 Introduction to Formaldehyde. *Formaldehyde: Exposure, Toxicity and Health Effects: The Royal Society of Chemistry*, 2018, p. 1-19.
[2] Kalapos, MP. A possible evolutionary role of formaldehyde. *Experimental & Molecular Medicine.*, 1999, 31(1), 1-4.
[3] Kalapos, MP. A possible evolutionary role of formaldehyde. *Exp Mol Med.*, 1999, 31(1), 1-4.

[4] Kebukawa, Y; David Kilcoyne, AL; Cody, GD. *Exploring the potential formation of organic soilds in chrondites and comets and through polymab The Astrophysical Journal.*, 2013, 771(1), 19.

[5] Wang, Z; Kang, H; Liu, H; Zhang, S; Xia, C; Wang, Z; et al. Dual-Network Nanocross-linking Strategy to Improve Bulk Mechanical and Water-Resistant Adhesion Properties of Biobased Wood Adhesives. *ACS Sustainable Chemistry & Engineering.*, 2020, 8(44), 16430-40.

[6] Gupta, SK. KAhdo----. Polymerization with Formaldehyde. In: *Reaction Engineering of Step Growth Polymerization. The Plenum Chemical Engineering Series.* Springer, Boston, MA., 1987.

[7] Desai, C. Meyler's side effects of drugs: The international encyclopedia of adverse drug reactions and interactions. *Indian J Pharmacol.*, 2016, 48(2), 224-.

[8] Swankie, W. editor *Effects of Temperature on the Emission Rate of Formaldehyde from Medium Density Fiberboard in a Controlled Chamber*, 2017.

[9] Liang, W; Yang, S; Yang, X. Long-Term Formaldehyde Emissions from Medium-Density Fiberboard in a Full-Scale Experimental Room: Emission Characteristics and the Effects of Temperature and Humidity. *Environmental Science & Technology.*, 2015, 49(17), 10349-56.

[10] Qin, D; Guo, B; Zhou, J; Cheng, H; Chen, X. Indoor air formaldehyde (HCHO) pollution of urban coach cabins. *Scientific Reports.*, 2020, 10.

[11] Qin, D; Guo, B; Zhou, J; Cheng, H; Chen, X. Indoor air formaldehyde (HCHO) pollution of urban coach cabins. *Scientific Reports.*, 2020, 10(1), 332.

[12] Norliana, S; Abdulamir, AS; Bakar, FA; Salleh, AB. The Health Risk of Formaldehyde to Human Beings. *American Journal of Pharmacology and Toxicology.*, 2009, 4(3).

[13] Jia, X; Jia, Q; Zhang, Z; Gao, W; Zhang, X; Niu, Y; et al. Effects of formaldehyde on lymphocyte subsets and cytokines in the peripheral blood of exposed workers. *PloS one.*, 2014, 9(8), e104069-e.

[14] Costa, S; García-Lestón, J; Coelho, M; Coelho, P; Costa, C; Silva, S; et al. Cytogenetic and Immunological Effects Associated with Occupational Formaldehyde Exposure. *Journal of Toxicology and Environmental Health, Part A.*, 2013, 76(4-5), 217-29.

[15] Sapmaz, HI; Sarsılmaz, M; Gödekmerdan, A; Ögetürk, M; Taş, U; Köse, E. Effects of formaldehyde inhalation on humoral immunity and protective effect of Nigella sativa oil: An experimental study. *Toxicology and Industrial Health.*, 2015, 32(9), 1564-9.

[16] Park, J; Yang, HS; Song, MK; Kim, DI; Lee, K. Formaldehyde exposure induces regulatory T cell-mediated immunosuppression via calcineurin-NFAT signalling pathway. *Scientific Reports.*, 2020, 10(1), 17023.

[17] Songur, A; Ozen, OA; Sarsilmaz, M. The Toxic Effects of Formaldehyde on the Nervous System. In: Whitacre DM, editor. *Reviews of Environmental Contamination and Toxicology*. New York, NY: Springer New York, 2010, p. 105-18.

[18] Tulpule, K; Dringen, R. Formaldehyde in brain: an overlooked player in neurodegeneration? *Journal of Neurochemistry.*, 2013, 127(1), 7-21.

[19] Arici, S; Karaman, S; Dogru, S; Cayli, S; Arici, A; Suren, M; et al. Central nervous system toxicity after acute oral formaldehyde exposure in rabbits: An experimental study. *Human & Experimental Toxicology.*, 2014, 33(11), 1141-9.

[20] Perna, RB; Bordini, EJ; Deinzer-Lifrak, M. A Case of Claimed Persistent Neuropsychological Sequelae of Chronic Formaldehyde Exposure: Clinical, Psychometric, and Functional Findings. *Archives of Clinical Neuropsychology.*, 2001, 16(1), 33-44.

[21] Bian, RX; Han, JY; Kim, JK; Choi, IS; Lee, SG; Park, JS; et al. The effect of chronic formaldehyde exposure on the hippocampus in chronic cerebral hypoperfusion rat model. *Toxicological & Environmental Chemistry.*, 2012, 94(6), 1211-24.

[22] Sorg, BA; Willis, JR; See, RE; Hopkins, B; Westberg, HH. Repeated Low-Level Formaldehyde Exposure Produces Cross-Sensitization to

Cocaine: Possible Relevance to Chemical Sensitivity in Humans. *Neuropsychopharmacology.*, 1998, 18(5), 385-94.

[23] Matsuoka, T; Takaki, A; Ohtaki, H; Shioda, S. Early changes to oxidative stress levels following exposure to formaldehyde in ICR mice. *The Journal of toxicological sciences.*, 2010, 35(5), 721-30.

[24] Kum, C; Kiral, F; Sekkin, S; Seyrek, K; Boyacioglu, M. Effects of Xylene and Formaldehyde Inhalations on Oxidative Stress in Adult and Developing Rats Livers. *Experimental Animals.*, 2007, 56(1), 35-42.

[25] Saito, Y; Nishio, K; Yoshida, Y; Niki, E. Cytotoxic effect of formaldehyde with free radicals via increment of cellular reactive oxygen species. *Toxicology.*, 2005, 210(2-3), 235-45.

[26] Squillacioti, G; Bellisario, V; Grosso, A; Ghelli, F; Piccioni, P; Grignani, E; et al. Formaldehyde, Oxidative Stress, and FeNO in Traffic Police Officers Working in Two Cities of Northern Italy. *International Journal of Environmental Research and Public Health.*, 2020, 17(5), 1655.

[27] Augenreich, M; Stickford, J; Stute, N; Koontz, L; Cope, J; Bennett, C; et al. Vascular dysfunction and oxidative stress caused by acute formaldehyde exposure in female adults. *American Journal of Physiology-Heart and Circulatory Physiology.*, 2020, 319(6), H1369-H79.

[28] Amiri, A; Turner-Henson, A. The Roles of Formaldehyde Exposure and Oxidative Stress in Fetal Growth in the Second Trimester. *Journal of Obstetric, Gynecologic & Neonatal Nursing.*, 2017, 46(1), 51-62.

[29] Leso, V; Macrini, MC; Russo, F; Iavicoli, I. Formaldehyde Exposure and Epigenetic Effects: A Systematic Review. *Applied Sciences.*, 2020, 10(7), 2319.

[30] Blackwell, M; Kang, H; Thomas, A; Infante, P. Formaldehyde: evidence of carcinogenicity. *American Industrial Hygiene Association journal.*, 1981, 42(7), A34, A6, A8, passim.

[31] McLaughlin, JK. Formaldehyde and cancer: a critical review. *International Archives of Occupational and Environmental Health.*, 1994, 66(5), 295-301.

[32] Andersen, ME; Gentry, PR; Swenberg, JA; Mundt, KA; White, KW; Thompson, C; et al. Considerations for refining the risk assessment process for formaldehyde: Results from an interdisciplinary workshop. *Regulatory Toxicology and Pharmacology.*, 2019, 106, 210-23.

[33] Speit, G; Schmid, O; Neuss, S; Schütz, P. Genotoxic effects of formaldehyde in the human lung cell line A549 and in primary human nasal epithelial cells. *Environmental and molecular mutagenesis.*, 2008, 49(4), 300-7.

[34] Viegas, S; Ladeira, C; Nunes, C; Malta-Vacas, J; Gomes, M; Brito, M; et al. Genotoxic effects in occupational exposure to formaldehyde: A study in anatomy and pathology laboratories and formaldehyde-resins production. *Journal of Occupational Medicine and Toxicology.*, 2010, 5(1), 25.

[35] Kawanishi, M; Matsuda, T; Yagi, T. Genotoxicity of formaldehyde: molecular basis of DNA damage and mutation. *Frontiers in Environmental Science.*, 2014, 2(36).

[36] Speit, G; Schmid, O. Local genotoxic effects of formaldehyde in humans measured by the micronucleus test with exfoliated epithelial cells. *Mutation Research/Reviews in Mutation Research.*, 2006, 613(1), 1-9.

[37] Peteffi, GP; da Silva, LB; Antunes, MV; Wilhelm, C; Valandro, ET; Glaeser, J; et al. Evaluation of genotoxicity in workers exposed to low levels of formaldehyde in a furniture manufacturing facility. *Toxicology and Industrial Health.*, 2016, 32(10), 1763-73.

[38] Costa, S; Pina, C; Coelho, P; Costa, C; Silva, S; Porto, B; et al. Occupational Exposure to Formaldehyde: Genotoxic Risk Evaluation By Comet Assay And Micronucleus Test Using Human Peripheral Lymphocytes. *Journal of Toxicology and Environmental Health, Part A.*, 2011, 74(15-16), 1040-51.

[39] Speit, G. The Implausibility of Systemic Genotoxic Effects Measured by the Comet Assay in Rats Exposed to Formaldehyde. *Journal of Proteome Research.*, 2006, 5(10), 2523-4.

[40] Kitaeva, LV; Mikheeva, EA; Shelomova, LF; Shvartsman, P. [Genotoxic effect of formaldehyde in somatic human cells *in vivo*]. *Genetika.*, 1996, 32(9), 1287-90.

[41] Duong, A; Steinmaus, C; McHale, CM; Vaughan, CP; Zhang, L. Reproductive and developmental toxicity of formaldehyde: A systematic review. *Mutation Research/Reviews in Mutation Research.*, 2011, 728(3), 118-38.

[42] Wang, Hx; Li, Hc; Lv, Mq; Zhou, Dx; Bai, Lz; Du, Lz; et al. Associations between occupation exposure to Formaldehyde and semen quality, a primary study. *Scientific Reports.*, 2015, 5(1), 15874.

[43] Wang, Hx; Zhou, Dx; Zheng, Lr; Zhang, J; Huo, Yw; Tian, H; et al. Effects of Paternal Occupation Exposure to Formaldehyde on Reproductive Outcomes. *Journal of Occupational and Environmental Medicine.*, 2012, 54(5), 518-24.

[44] Özen, OA; Akpolat, N; Songur, A; Kuş, İ; Zararsiz, İ; Özaçmak, VH; et al. Effect of formaldehyde inhalation on Hsp70 in seminiferous tubules of rat testes: an immunohistochemical study. *Toxicology and Industrial Health.*, 2005, 21(9), 249-54.

[45] Xu, W; Zhang, W; Zhang, X; Dong, T; Zeng, H; Fan, Q. Association between formaldehyde exposure and miscarriage in Chinese women. *Medicine.*, 2017, 96(26), e7146.

[46] CHAPTER 12 Reproductive and Developmental Toxicity of Formaldehyde Exposure in Humans. *Formaldehyde: Exposure, Toxicity and Health Effects: The Royal Society of Chemistry*, 2018, p. 265-92.

[47] Pidoux, G; Gerbaud, P; Guibourdenche, J; Thérond, P; Ferreira, F; Simasotchi, C; et al. Formaldehyde Crosses the Human Placenta and Affects Human Trophoblast Differentiation and Hormonal Functions. *PLOS ONE.*, 2015, 10(7), e0133506.

[48] Mathur, N; Rastogi, SK. Respiratory effects due to occupational exposure to formaldehyde: Systematic review with meta-analysis. *Indian J Occup Environ Med.*, 2007, 11(1), 26-31.

[49] Zhang, J; Sun, R; Chen, Y; Tan, K; Wei, H; Yin, L; et al. Small molecule metabolite biomarker candidates in urine from mice exposed to formaldehyde. *International journal of molecular sciences.*, 2014, 15(9), 16458-68.

[50] Fraenkel-Conrat, H; Olcott, HS. The Reaction of Formaldehyde with Proteins. V. Cross-linking between Amino and Primary Amide or Guanidyl Groups. *Journal of the American Chemical Society.*, 1948, 70(8), 2673-84.

[51] Kamps, JJAG; Hopkinson, RJ; Schofield, CJ; Claridge, TDW. How formaldehyde reacts with amino acids. *Communications Chemistry.*, 2019, 2(1), 126.

[52] Metz, B; Kersten, GFA; Hoogerhout, P; Brugghe, HF; Timmermans, HAM; de Jong, A; et al. Identification of Formaldehyde-induced Modifications in Proteins: Reactions With Model Peptides*. *Journal of Biological Chemistry.*, 2004, 279(8), 6235-43.

[53] Tayri-Wilk, T; Slavin, M; Zamel, J; Blass, A; Cohen, S; Motzik, A; et al. Mass spectrometry reveals the chemistry of formaldehyde cross-linking in structured proteins. *Nature Communications.*, 2020, 11(1), 3128.

[54] Metz, B; Kersten, GF; Hoogerhout, P; Brugghe, HF; Timmermans, HA; de Jong, A; et al. Identification of formaldehyde-induced modifications in proteins: reactions with model peptides. *The Journal of biological chemistry.*, 2004, 279(8), 6235-43.

[55] Chen, NH; Djoko, KY; Veyrier, FJ; McEwan, AG. Formaldehyde Stress Responses in Bacterial Pathogens. *Frontiers in Microbiology.*, 2016, 7(257).

[56] Nie, CL; Wei, Y; Chen, X; Liu, YY; Dui, W; Liu, Y; et al. Formaldehyde at Low Concentration Induces Protein Tau into Globular Amyloid-Like Aggregates *In Vitro* and *In Vivo*. *PLOS ONE.*, 2007, 2(7), e629.

[57] Toby, S; Rutz, F. *Polymerization of gaseous formaldehyde Journal of Polymer Science.*, 1962, 60(169).

[58] Papon, P; Leblond, J; Meijer, PHE. Phase Transitions in Liquids and Solids: Solidification and Melting. *The Physics of Phase Transitions: Concepts and Applications.* Berlin, Heidelberg: Springer Berlin Heidelberg, 2002, p. 79-122.

[59] Brown, Northrop; Funck, Dennis L; Earle, SC. *Polymerization of aqueous formaldehyde to produce high molecular weight polyoxymethylene.* US3000861A, United States https://patents.google.com/patent/US3000861A/en. 1957.

[60] Henri, J. Vapor phase polymerization of formaldehyde, US3000861A, United States https://patents.google.com/patent/US3091599. 1963.

[61] McGill, PR; Söhnel, T. A study of gas phase and surface formaldehyde polymerisation from first principles. *Physical Chemistry Chemical Physics.*, 2012, 14(2), 858-68.

[62] Bachmann, MO; Myers, JE. Influences on sick building syndrome symptoms in three buildings. *Social science & medicine*, (1982), 1995, 40(2), 245-51.

[63] Huang, KC; Tsay, YS; Lin, FM; Lee, CC; Chang, JW. Efficiency and performance tests of the sorptive building materials that reduce indoor formaldehyde concentrations. *PLOS ONE.*, 2019, 14(1), e0210416.

[64] Vikrant, K; Cho, M; Khan, A; Kim, KH; Ahn, WS; Kwon, EE. Adsorption properties of advanced functional materials against gaseous formaldehyde. *Environmental Research.*, 2019, 178, 108672.

[65] Lara-Ibeas, I; Megías-Sayago, C; Louis, B; Le Calvé, S. Adsorptive removal of gaseous formaldehyde at realistic concentrations. *Journal of Environmental Chemical Engineering.*, 2020, 8(4), 103986.

[66] Junyi, W; Yousheng, T. Removal of Formaldehyde from the Indoor Environment Using Porous Carbons and Silicas. *Recent Innovations in Chemical Engineering.*, 2020, 13(3), 194-202.

[67] Carter, EM; Katz, LE; Speitel, GE; Ramirez, D. Gas-Phase Formaldehyde Adsorption Isotherm Studies on Activated Carbon:

Correlations of Adsorption Capacity to Surface Functional Group Density. *Environmental Science & Technology.*, 2011, 45(15), 6498-503.

[68] Im, DH; Chu, YS; Song, H; Lee, JK. Formaldehyde Adsorption and Physical Characteristics of Hydrothermal Reacted Panels Using Porous Materials. *J Korean Ceram Soc.*, 2009, 46(6), 627-0.

[69] Krishnamurthy, A; Thakkar, H; Rownaghi, AA; Rezaei, F. Adsorptive Removal of Formaldehyde from Air Using Mixed-Metal Oxides. *Industrial & Engineering Chemistry Research.*, 2018, 57(38), 12916-25.

[70] Vikrant, K; Cho, M; Khan, A; Kim, KH; Ahn, WS; Kwon, EE. Adsorption properties of advanced functional materials against gaseous formaldehyde. *Environmental research.*, 2019, 178, 108672.

[71] Srisuda, S; Virote, B. Adsorption of formaldehyde vapor by amine-functionalized mesoporous silica materials. *J Environ Sci* (China)., 2008, 20(3), 379-84.

[72] Kibanova, D; Sleiman, M; Cervini-Silva, J; Destaillats, H. Adsorption and photocatalytic oxidation of formaldehyde on a clay-TiO2 composite. *Journal of Hazardous Materials.*, 2012, 233-9.

[73] Liu, L; Zhang, D; Liu, J. Computer aided design of water-resistant adsorbent for formaldehyde abatement. *IOP Conference Series: Materials Science and Engineering.*, 2019, 609, 042108.

[74] Liu, L; Liu, J; Zeng, Y; Tan, SJ; Do, DD; Nicholson, D. Formaldehyde adsorption in carbon nanopores – New insights from molecular simulation. *Chemical Engineering Journal.*, 2019, 370, 866-74.

[75] Su, C; Liu, K; Guo, Y; Li, H; Zeng, Z; Li, L. The role of pore structure and nitrogen surface groups in the adsorption behavior of formaldehyde on resin-based carbons. *Surface and Interface Analysis.*, 2021, 53(3), 330-9.

[76] Tang, X; Bai, Y; Duong, A; Smith, MT; Li, L; Zhang, L. Formaldehyde in China: production, consumption, exposure levels, and health effects. *Environment international.*, 2009, 35(8), 1210-24.

[77] World Health Organization, *WHO guidelines for indoor air quality: selected pollutants*, WHO Regional Office for Europe, Copenhagen, 2010.
[78] Steinhauser, MO; Hiermaier, S. A Review of Computational Methods in Materials Science: Examples from Shock-Wave and Polymer Physics. *International journal of molecular sciences.*, 2009, 10(12), 5135-216.
[79] Bricker, CE; Johnson, HR. Spectrophotometric Method for Determining Formaldehyde. *Industrial & Engineering Chemistry Analytical Edition.*, 1945, 17(6), 400-2.
[80] West, PW; Sen, B. Spectrophotometric determination of traces of formaldehyde. *Fresenius' Zeitschrift für analytische Chemie.*, 1956, 153(3), 177-83.
[81] Sawicki, E; Hauser, TR; McPherson, S. Spectrophotometric Determination of Formaldehyde and Formaldehyde-Releasing Compounds with Chromotropic Acid, 6-Amino-1-naphthol-3-sulfonic Acid (J Acid), and 6-Anilino-1-napthol-3-sulfonic Acid (Phenyl J Acid). *Analytical Chemistry.*, 1962, 34(11), 1460-4.
[82] Hladová, M; Martinka, J; Rantuch, P; Nečas, A. Review of Spectrophotometric Methods for Determination of Formaldehyde. *Research Papers Faculty of Materials Science and Technology Slovak University of Technology.*, 2019, 27(44), 105-20.
[83] Gasparini, F; Weinert, PL; Lima, LS; Pezza, L; Pezza, HR. A simple and green analytical method for the determination of formaldehyde. *Journal of the Brazilian Chemical Society.*, 2008, 19, 1531-7.
[84] Yue, X; Zhang, Y; Xing, W; Chen, Y; Mu, C; Miao, Z; et al. A Sensitive and Rapid Method for Detecting Formaldehyde in Brain Tissues. *Analytical Cellular Pathology.*, 2017, 2017, 9043134.
[85] Lamarca, RS; Luchiari, NdC; Bonjorno, AF; Passaretti Filho, J; Cardoso, AA; Lima Gomes, PCFd. Determination of formaldehyde in cosmetic products using gas-diffusion microextraction coupled with a smartphone reader. *Analytical Methods.*, 2019, 11(29), 3697-705.

[86] Khare, VSS. Nanobiomimicry at quantum scales: synthetic imperfections to imitating low-dimensional biomimetic polymer confinement of TiO2 at self-assembled heterogeneous interfaces. In: Hong NH (ed) *Nano-sized multifunctional materials*. Elsevier, Amsterdam, pp. 231-270, 2019.

[87] Nengsih, S; Umar, AA; Salleh, MM; Oyama, M. Detection of Formaldehyde in Water: A Shape-Effect on the Plasmonic Sensing Properties of the Gold Nanoparticles. *Sensors.*, 2012, 12(8), 10309-25.

[88] Wang, D; Li, Z; Zhou, J; Fang, H; He, X; Jena, P; et al. Simultaneous Detection and Removal of Formaldehyde at Room Temperature: Janus Au@ZnO@ZIF-8 Nanoparticles. *Nano-Micro Letters.*, 2017, 10(1), 4.

[89] Qian, X. The effect of cooperativity on hydrogen bonding interactions in native cellulose Iβ from ab initio molecular dynamics simulations. *Molecular Simulation.*, 2008, 34(2), 183-91.

[90] Xiang, DL; Hou, SM; Tong, DG. Amorphous Eu0.9Ni0.1B6 Nanoparticles for Formaldehyde Vapor Detection. *ACS Applied Nano Materials.*, 2019, 2(7), 4048-52.

[91] Feng, L; Musto, CJ; Suslick, KS. A Simple and Highly Sensitive Colorimetric Detection Method for Gaseous Formaldehyde. *Journal of the American Chemical Society.*, 2010, 132(12), 4046-7.

[92] Chang, X; Wang, Z; Qi, Y; Kang, R; Cui, X; Shang, C; et al. Dynamic Chemistry-Based Sensing: A Molecular System for Detection of Saccharide, Formaldehyde, and the Silver Ion. *Analytical Chemistry.*, 2017, 89(17), 9360-7.

[93] Li, P; Zhang, D; Zhang, Y; Lu, W; Wang, W; Chen, T. Ultrafast and Efficient Detection of Formaldehyde in Aqueous Solutions Using Chitosan-based Fluorescent Polymers. *ACS Sensors.*, 2018, 3(11), 2394-401.

[94] Tang, Y; Kong, X; Xu, A; Dong, B; Lin, W. Development of a Two-Photon Fluorescent Probe for Imaging of Endogenous Formaldehyde in Living Tissues. *Angewandte Chemie* (International ed in English)., 2016, 55(10), 3356-9.

[95] Rao, Y; Comstock, M; Eisenthal, KB. Absolute Orientation of Molecules at Interfaces. *The Journal of Physical Chemistry B.*, 2006, 110(4), 1727-32.

[96] Zhang, Y; Xie, LZ; Yuan, CX; Zhang, CL; Liu, S; Peng, YQ; et al. A ppb-Level Formaldehyde Gas Sensor Based on Rose-Like Nickel Oxide Nanoparticles Prepared Using Electrodeposition Process. *Nano.*, 2016, 11(01), 1650009.

[97] Castro-Hurtado, I; Herrán, J; Mandayo, GG; Castaño, E. Studies of influence of structural properties and thickness of NiO thin films on formaldehyde detection. *Thin Solid Films.*, 2011, 520(3), 947-52.

[98] Kannan, PK; Saraswathi, R. An impedance sensor for the detection of formaldehyde vapor using ZnO nanoparticles. *Journal of Materials Research.*, 2017, 32(14), 2800-9.

[99] Bai, Z; Li, S; Xu, J; Zhou, Y; Gu, S; Tao, Y; et al. Fabrication and gas-sensing properties of hierarchical ZnO replica using down as template. *Applied Physics A.*, 2016, 122(6), 622.

[100] Ehsan, MA; Rehman, A. Facile and scalable fabrication of nanostructured nickel thin film electrodes for electrochemical detection of formaldehyde. *Analytical Methods.*, 2020, 12(32), 4028-36.

[101] Lahem, D; Lontio, FR; Delcorte, A; Bilteryst, L; Debliquy, M. Formaldehyde gas sensor based on nanostructured nickel oxide and the microstructure effects on its response. *IOP Conference Series: Materials Science and Engineering.*, 2016, 108, 012002.

[102] Gao, L; Fu, H; Zhu, J; Wang, J; Chen, Y; Liu, H. Synthesis of SnO2 nanoparticles for formaldehyde detection with high sensitivity and good selectivity. *Journal of Materials Research.*, 2020, 35(16), 2208-17.

[103] Choi, D; Sonkaria, S; Fox, SJ; Poudel, S; Kim, Sy; Kang, S; et al. Quantum scale biomimicry of low dimensional growth: An unusual complex amorphous precursor route to TiO2 band confinement by shape adaptive biopolymer-like flexibility for energy applications. *Scientific Reports.*, 2019, 9(1), 18721.

[104] Wu, K; Kong, XY; Xiao, K; Wei, Y; Zhu, C; Zhou, R; et al. Engineered Smart Gating Nanochannels for High Performance in Formaldehyde Detection and Removal. *Advanced Functional Materials.*, 2019, 29(14), 1807953.

[105] Salthammer, T. Formaldehyde sources, formaldehyde concentrations and air exchange rates in European housings. *Building and Environment.*, 2019, 150, 219-32.

In: A Comprehensive Guide to Formaldehyde ISBN: 978-1-53619-465-4
Editor: Natasja A. Bach © 2021 Nova Science Publishers, Inc.

Chapter 3

ELECTROCHEMICAL ASSESSMENT OF FORMALDEHYDE WITH DOPED NANOROD MATERIALS

Mohammed Muzibur Rahman[1,2,*], *Abdullah M. Asiri*[1,2], *M. M. Alam*[3] *and Jamal Uddin*[4]

[1]Department of Chemistry, Faculty of Science, King Abdulaziz University, Jeddah, Saudi Arabia
[2]Center of Excellence for Advanced Materials Research, King Abdulaziz University, Jeddah, Saudi Arabia
[3]Department of Chemical Engineering and Polymer Science, Shahjalal University of Science and Technology, Sylhet, Bangladesh
[4]Center for Nanotechnology, Department of Natural Sciences, Coppin State University, Baltimore, MD, US

---

[*] Corresponding Author's E-mail: mmrahman@kau.edu.sa.

ABSTRACT

In this approach, the growth and development of doped nanostructure materials (i.e., nanorods) by using hydrothermal method in alkaline phase were discussed. The structural, optical and chemical properties of nanorods (NRs) were characterized using various methods such as UV/vis., FTIR, Raman spectroscopy, powder XRD, and FESEM, XEDS, XPS, etc. Doped $ZnFe_2O_4$ is an attractive nanorod for potential application in chemical sensing by easy and reliable electrochemical method, where formaldehyde is considered as a model compound. The chemical sensor performances are exhibited the higher sensitivity, good stability, and repeatability of the sensor enhanced significantly using doped NRs of thin-film with conducting coating binders on silver electrodes. The calibration plot is linear over the large dynamic range, where the sensitivity and detection limit were calculated based on signal/noise ratio ($\sim^{3N}/_S$) in short response time. At last, it is concluded that the prepared doped nanomaterial could be implemented in a broad-scale for an efficient electro-chemical sensor applications for environmental and healthcare fields.

LITERATURE REVIEW

Generally, during the last few decades, semiconductor doped nano-materials have received great attention owing to their electronic, magnetic, electrical, optoelectronic, mechanical properties and their prospective applications in various fields. Spinel material might be a promising candidate because of their high specific surface-area, low-resistance, fascinating electrochemical and optical behaviors [1-3]. Zinc ferrites are also technologically significant doped nanomaterials due to their exceptional mechanical, electrical, thermal, and magnetic characteristics. Recently, it has gained much attention in ferrites (iron doped-semiconductor materials) in order to improve their physical properties and expand their potential applications [4-6]. It has not only studied the fundamental of magnetism, but also has enormous prospective in technological aspects (i.e., magnetic materials, sensors, catalysts, and absorbent materials) [7-14]. Recently, several articles are reported on the basis of $ZnFe_2O_4$ nanomaterials preparation and studied the magnetic properties only [15, 16]. Herein, it is synthesized spinel $ZnFe_2O_4$ nanorod by simple, facile, economical, non-toxic, reproducible, and reliable low-temperature hydrothermal method. The structure and morphology of the calcined $ZnFe_2O_4$ nanorods were investigated and applied for the development of highly sensitive formaldehyde chemical sensor at room conditions. Generally, chemical sensing investigation have been explored with the metal oxide nanostructures for the detection and quantification of various chemicals such as methanol, acetone, phenyl hydrazine, ethanol, chloroform, dichloromethane, etc., which are not environmentally safe [17, 18]. The sensing mechanism with doped metal oxides thin films utilized mainly the properties of porous film formed by the physi-sorption and chemi-sorptions techniques. The hazardous chemical detection is based on the current responses of the fabricated thin films, which caused by the presence of chemical components in the reacting system in aqueous phase [19-21]. The key attempts are focused on detecting the minimum amount of formaldehyde necessary for the fabricated spinel $ZnFe_2O_4$ NRs sensors for electrochemical investigation.

In this study, it is synthesized the spinel $ZnFe_2O_4$ NRs by hydrothermal technique with almost controlled rod-shape structure, which displayed a continuous structural and morphological improvement in transition-metal doped semiconductor materials. The spinel $ZnFe_2O_4$ NR allows very sensitive recognition and transduction in the chemical interaction to change the electrochemical properties. Finally, spinel $ZnFe_2O_4$ NR is fabricated to make a simple, reliable and efficient chemical sensor onto a side-polished silver electrode surfaces and executed the chemical sensing performances with toxic and carcinogenic formaldehyde chemical at room conditions. To best of our knowledge, this is the first report for highly sensitive detection of formaldehyde with spinel $ZnFe_2O_4$ NRs using simple and reliable electrochemical method in short response time.

EXPERIMENTAL SECTIONS

Chemical Reagents and Apparatus

Formaldehyde, Ferric chloride ($FeCl_3$), Zinc chloride ($ZnCl_2$), Ethyl acetate, Disodium phosphate, Butyl carbitol acetate, Ammonia hydroxide (25%), Monosodium phosphate, and all other chemicals were used in analytical grade and purchased from Sigma-Aldrich Company. Stock solution of formaldehyde (10.0 M) was prepared in double distilled water. The spinel $ZnFe_2O_4$ NR was examined with UV/visible spectroscopy (Lamda-950, Perkin Elmer, Germany). FT-IR spectra were measured for calcined $ZnFe_2O_4$ NRs with a spectrophotometer (Spectrum-100 FT-IR) in the mid-IR range, which was obtained from Perkin Elmer, Germany. The powder X-ray diffraction (XRD) prototypes were assessed with X-ray diffractometer (XRD; X'Pert Explorer, PANalytical diffractometer) prepared with $CuK_\alpha 1$ radiation ($\lambda = 1.5406$ nm) using a generator voltage of 40.0 kV and current of 35.0 mA applied for the measurement. Morphology of spinel $ZnFe_2O_4$ NR was examined on FE-SEM instrument (FESEM; JSM-7600F, Japan). Elemental analysis of spinel $ZnFe_2O_4$ NR was investigated using EDS from JEOL, Japan. Electrochemical method was employed by Electrometer (Kethley, 6517A, Electrometer, USA) in room condition.

Synthesis and Growth Mechanism of Spinel $ZnFe_2O_4$ NRs

Large-scale production of spinel $ZnFe_2O_4$ NR was prepared by hydrothermal method at low-temperature by using zinc chloride ($ZnCl_3$), iron chloride ($FeCl_3$), and ammonium hydroxide (NH_4OH). In a typical reaction process, 0.1 M $ZnCl_3$ dissolved in 50.0 ml deionized (DI) water mixed with 50.0 ml DI solution of 0.1 M $FeCl_3$ under continuous stirring.

The pH of the solution was adjusted to 10.25 by the addition of NH_4OH and resulting mixture was shacked and stirred continuously for 10.0 minutes at room condition. After stirring, the mixture was then put into Teflon auto-clave and transferred in the oven for 16.0 hours. The temperature of oven was controlled manually throughout the reaction process at 150°C. After heating the reactant mixtures, the Teflon-auto clave was kept for cooling at room condition until it reached to room temperature. The final products were achieved, which was washed with DI water, ethanol, and acetone for several times subsequently and dried at room-temperature for structural and optical characterization. The growth mechanism of spinel $ZnFe_2O_4$ NR can be apprehended on the basis of chemical reactions and nucleation, as well as growth of $ZnFe_2O_4$ crystals. The probable reaction mechanisms are proposed for achieving the doped spinal nanomaterial oxides, which are appended below.

$$ZnCl_2(s) \rightarrow Zn^{2+}_{(aq)} + 2Cl^-_{(aq)} \tag{1}$$

$$FeCl_3(s) \rightarrow Fe^{3+}_{(aq)} + 3Cl^-_{(aq)} \tag{2}$$

$$NH_3 + H_2O \leftrightarrow NH_4^+ + OH^- \tag{3}$$

$$Fe^{3+}_{(aq)} + Zn^{2+}_{(aq)} + 8OH^-_{(aq)} \rightarrow ZnFe_2O_{4(s)} + 4H_2O \tag{4}$$

The precursors of $ZnCl_3$ and $FeCl_3$ are soluble in alkaline medium (NH_4OH reagent) according to the equation of (1)-(3). After addition of NH_4OH into the mixture of metal chlorides solution, it was strongly stirred for few minutes at room temperature. The reaction is enhanced gradually according to the equation (4). Then the solution was washed thoroughly with acetone and kept for drying at room temperature. During the total synthesis process, NH_4OH acts a pH buffer to control the pH value of the solution and slow donate of OH^- ions.

When the concentrations of the Zn^{2+}, Fe^{3+}, and OH^- ions are reached above the critical value, the precipitation of $ZnFe_2O_4$ nuclei instigate to start. As there is high concentration of Zn^{2+} ion in the solution, the nucleation of $ZnFe_2O_4$ crystals become easier owing to the lower activation energy barrier of heterogeneous nucleation. However, as the concentration of Fe^{3+} subsistence, a number of larger $ZnFe_2O_4$ crystals with a rod-shape morphology form in nano-level. The shape of calcined $ZnFe_2O_4$ NRs is approximately reliable with the growth pattern of spinel crystals [22, 23]. Finally, the as-grown $ZnFe_2O_4$ nanostructure products were calcined at 400°C for 5 hours in the furnace (Barnstead Thermolyne, 6000 Furnace, USA). The calcined products were characterized in detail in terms of their morphological, structural, optical properties, and applied for formaldehyde chemical sensing.

Fabrication and Detection of Formaldehyde Using Spinel $ZnFe_2O_4$ NRs

Phosphate buffer solution (PBS, 0.1M, pH 7.0) is made by adding of 0.2M Na_2HPO_4 and 0.2M NaH_2PO_4 solution in 100.0 mL de-ionize water. Silver electrode (AgE, surface area, 0.0216 cm^2) is fabricated by spinel $ZnFe_2O_4$ nanorods with butyl carbitol acetate (BCA) and ethyl acetate (EA) as a conducting binding agent. Then it is kept into the oven at 65°C for 12 hours until the film is completely uniform and dry. An electrochemical cell is constructed with spinel $ZnFe_2O_4$ NR coated AgE as a working electrode and Pd wire is used a counter electrode. As received formaldehyde (10.0M) chemical is diluted at different concentrations with DI water and used as a target chemical. Amount of 0.1M PBS is kept constant in the beaker as 10.0 mL throughout the total chemical analysis. Analyte solution is arranged with various concentrations of formaldehyde from 0.01 µM to 10.0M. The sensitivity is calculated from the slope of voltage versus current from the calibration plot. Electrometer is used as a voltage sources for electrochemical method in two electrodes system.

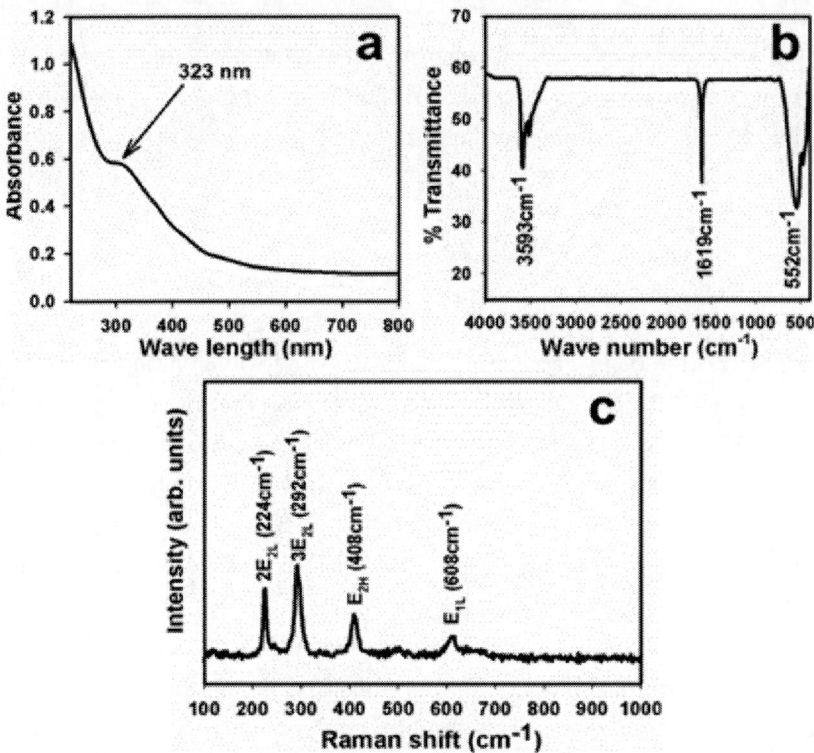

Figure 1. Optical evaluation of material. (a) UV/visible, (b) FT-IR, and (c) Raman spectroscopy of spinel $ZnFe_2O_4$ NRs at room conditions.

RESULTS AND DISCUSSIONS

Investigation of Optical Properties

The optical property of the spinel $ZnFe_2O_4$ nanorod is one of the important features for the evaluation of its photocatalytic action. Accordingly, optical properties of the calcined $ZnFe_2O_4$ nanorod were revealed by using UV/visible spectrophotometer and presented in Figure 1a. UV/visible absorption spectrum exhibited absorption peak 323 nm, which is corresponded the characteristic absorption peak. No extra peak associated with impurities and structural defects are observed in the optical spectrums, which proved that the synthesized nanorods are well crystalline of spinel $ZnFe_2O_4$. Band-gap energy is calculated on the sources of the highest absorption band of spinel $ZnFe_2O_4$ nanorods and obtained to be 3.8390 eV (E_{bg}), according to the following equation.

$$E_{bg} = \frac{1240}{\lambda} \text{ (eV)}$$

where E_{bg} is the band-gap energy and λ_{max} is the wavelength (323.0 nm) of the calcined $ZnFe_2O_4$ nanorods. It can been observed, the optical band-gap of calcined $ZnFe_2O_4$ nanorods is 3.8390, which is in good conformity with the doped Fe-doped ZnO band structure [24]. This proposed that the calcined $ZnFe_2O_4$ nanorods are transition-doped semiconductor with measured energy-value. E_{bg} is accredited to inter-band doped transition and considered as the accurate energy gap.

The spinel $ZnFe_2O_4$ nanorods are also exemplified from the atomic and molecular vibrations. To expect the motivated recognition, FT-IR spectra basically in the region of 400-4000 cm^{-1} are investigated. Figure 1b exhibits the FT-IR spectrum of the calcined $ZnFe_2O_4$. It presents several bands at 538, 552, 1619, and 3593 cm^{-1}. These observed vibration bands (552 cm^{-1} and shoulder peak at 538 cm^{-1}) could be assigned as metal-oxygen (Fe-O and Zn-O modes) stretching vibrations respectively [25], which verified the formation of doped semiconductor nanostructure materials. The additional observed vibration bands may be assigned to O-H stretching (3593 cm^{-1}) and O-H bending vibration (1619 cm^{-1}). The absorption bands at 1619 cm^{-1} and 3593 cm^{-1} normally exhibits from water, which usually semiconductor NRs absorbed from the environment due to their meso-porous structure. Finally, the observed vibration bands at low frequencies regions suggested the formation of spinel $ZnFe_2O_4$ NRs.

Raman spectroscopy is generally established and utilized in material chemistry, since the information is specific to the chemical bonds and symmetry of metal-oxygen stretching or vibrational modes. Usually, there are three vibration modes in ZnO nanomaterial crystal: A_1, E_1 and E_2, of which A_1 and E_1 split into longitudinal (A_{1L}, E_{1L}) and transverse (A_{1T}, E_{1T}) ones and E_2 contains low and high frequency phonons (E_{2L} and E_{2H}) [26]. For Fe-ZnO, the Raman spectrum keeps the same as pure ZnO. Spinel $ZnFe_2O_4$ NR is not significantly altered the Raman spectra as well as the crystal of ZnO nanostructure [27]. Here, Figure 2c confirms the Raman spectrum, where key aspects of the wave number are employed at about 224 cm^{-1} ($2E_{2L}$), 292 cm^{-1} ($3E_{2L}$), 408 cm^{-1} (E_{2H}), and 608 cm^{-1} (E_{1L}) for metal-oxygen (Fe-O and Zn-O) stretching vibrations. These large bands can be assigned to a cubic phase of spinel $ZnFe_2O_4$ NRs. At 608 cm^{-1}, higher wave-number shift is revealed owing to the different dimensional effects of the NRs.

Investigation of Structural Properties

The general morphologies of the spinel $ZnFe_2O_4$ NRs are examined using FE-SEM. High-resolution FE-SEM images of NRs are presented in Figure 2a and Figure 2b. The FE-SEM images are composed of nanostructure materials with cumulative structure in rod-shape and reveals that the NRs are grown in a very high density. The average diameter of spinel $ZnFe_2O_4$ NR is 210.0 nm in the range of 117.81 nm to 297.36 nm, which is calculated from Figure 2a. The nanostructures are arranged in such a special fashion that they are making a perfect rod-shaped morphology. It was interesting to see that the diameters of the calcined nanostructures are uniform along their cross-section (diameter). It is also suggested that almost all of the products possessed in aggregated spinel $ZnFe_2O_4$ NRs. Excitingly, it was also observed that the major portion of these rod-shaped structures have a perfectly hexagonal which makes the nanostructures a rod-like morphology (Figure 2(b)). When the size of doped material decreases into nanometer-sized scale, the surface area increases significantly, this improves the energy of the system, making the re-distribution of Zn and Fe ions in the crystalline phases. The nanometer-sized rods could have tightly packed into the lattice, which is good agreement with the previous results [28].

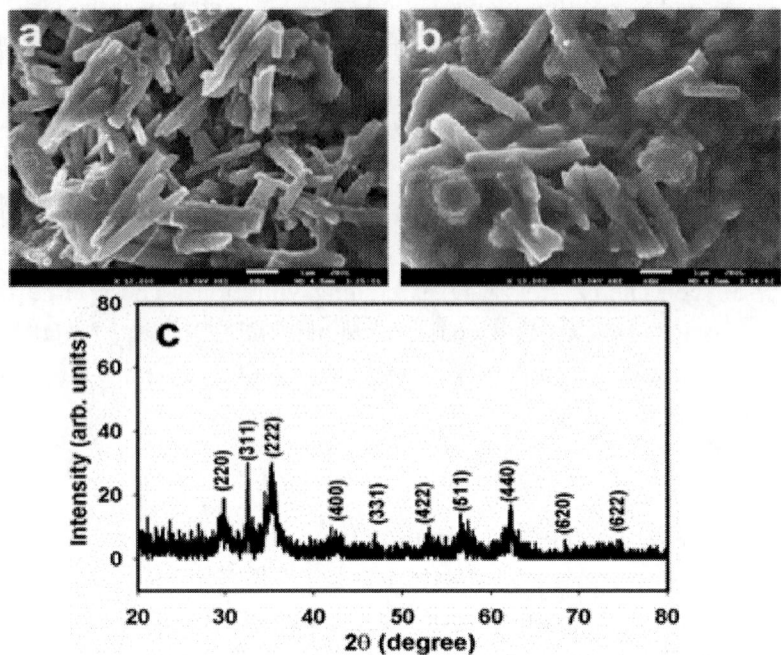

Figure 2. Morphological and structural analysis of material. (a-b) FE-SEM images and (b) Powder x-ray diffraction pattern of spinel $ZnFe_2O_4$ NRs at room conditions.

Crystallinity and crystal phase of the calcined $ZnFe_2O_4$ NR were investigated. X-ray diffraction patterns of spinel NR are represented in Figure 2c. The spinel $ZnFe_2O_4$ NR samples were checked and exhibited as face-centered cubic shapes called as Franklinite. The $ZnFe_2O_4$ sample was calcined at 400°C into furnace to start the formation of nano-crystalline phases. Figure 2c reveals characteristic crystallinity of the calcined $ZnFe_2O_4$ NR and their aggregative arrangement, which is investigated using powder X-ray crystallography. All the reflection peaks in this prototype were initiated to correspond with spinel $ZnFe_2O_4$ phase having face-centered cubic Franklinite geometry [JCPDS # 073-3824]. The phases demonstrated the key features with indices for crystalline $ZnFe_2O_4$ at various 2θ values of (220), (311), (222), (400), (331), (422), (511), (440), (620), and (622). The spinel $ZnFe_2O_4$ nanorods have a high degree of crystallinity. All of the peaks match well with Bragg reflections of the standard Franklinite structure. These confirmed that there is major number and amount of crystalline calcined $ZnFe_2O_4$ NR present in nano-rods [29]. No other apparent reflections were detected demonstrating in presence of a subsequent crystal phase.

Detection of Formaldehyde Using Spinel $ZnFe_2O_4$ NRs by Electrochemical Method

The potential application of calcined $ZnFe_2O_4$ NRs as a chemical sensor (formaldehyde detection) has been executed for measuring, detecting, and quantifying hazardous and carcinogenic chemicals, which are not environmentally safe. The NRs of $ZnFe_2O_4$ sensors have advantages such as stability in air, non-toxicity, not-carcinogenic, electrochemical activity, simplicity to accumulate, and bio-safe characteristics [30-34]. As in the case of formaldehyde chemical sensors, the phenomenon is that the current response in I-V technique using $ZnFe_2O_4$ considerably transforms when aqueous formaldehyde are adsorbed on the large surface area. The spinel $ZnFe_2O_4$ NRs are used for the modification of chemical sensor, where formaldehyde is considered as target analyte. The fabricated-surface of $ZnFe_2O_4$ NR sensor is equipped with conducting binders on the silver electrode surface, which is shown in the Figure 3a. The fabricated surface is transferred into the oven at low temperature to make it smooth, dry, stable, and uniform the surface entirely. Hypothetical electrochemical current responses of chemical sensor are prospected containing spinel $ZnFe_2O_4$ NR thin film as a function of current versus potential, which is presented in Figure 3b. The real-electrical signals of target formaldehyde are examined by easy and reliable electrochemical method using $ZnFe_2O_4$ NRs fabricated AgE surface, which is displayed in Figure 3c. The time delaying of electrometer was kept for 1 sec. A considerable intensification of the electrical signal with applied potential is perceptibly confirmed. The simple, reliable, probable reaction mechanism is presented in Figure 3d in presence of formaldehyde on fabricated spinel $ZnFe_2O_4$ NRs sensor surfaces by electrochemical technique. The formaldehyde is changed into water and carbon dioxide in presence of spinel $ZnFe_2O_4$ NRs ($2O^-$) by discharging electrons ($2e^-$) into the reaction system (conduction band, C. B.), which enhanced and improved the current signals against potential during electrochemical measurement at room conditions.

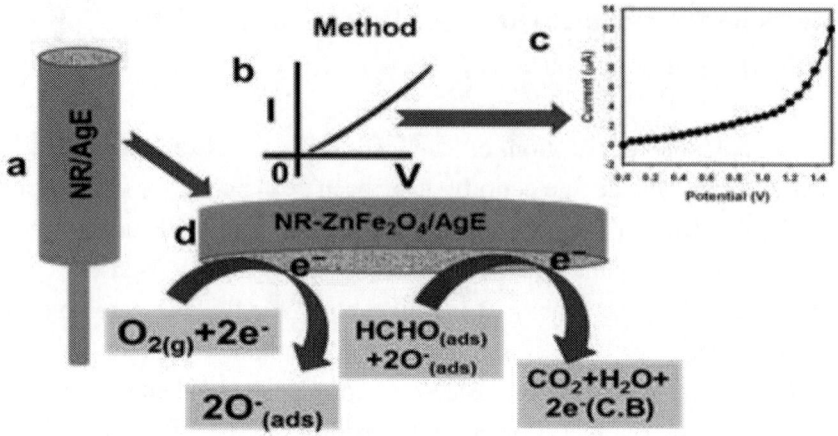

Figure 3. Schematic representation of (a) fabricated with NRs with conducting binders (EC and BCA), (b) Electrochemical detection methodology (theoretical), (c) outcomes of electrochemical experimental plot, (d) reaction mechanism of formaldehyde in presence of semiconductor spinel $ZnFe_2O_4$ NRs.

Figure 4a reveals the current responses without (gray-dotted) and with (dark-dotted) coating of spinel $ZnFe_2O_4$ NRs on AgE working electrode surfaces. With NRs fabricating surface, the current response is reduced compared to without fabricated surface, which indicates the surface is slightly inhibited with coated nanomaterials. The current changes for the nanomaterials coated film before (gray-dotted) and after (dark-dotted) injection of 50.0 µL formaldehyde (0.01 µM) in 10.0 mL PBS solution, which is displayed in Figure 4b. These significant changes of surface current are examined in every injection of the target formaldehyde into the bulk solution by electrometer. 10.0 mL of 0.1M PBS solution is initially taken into the cell and added the low to high concentration of formaldehyde drop-wise successively from the stock solution. This affinity obviously enhance the sensitivity of spinel $ZnFe_2O_4$ NRs towards formaldehyde sensing purposes, which may be owing to the fast electron communication and good electro-catalytic oxidation properties of transition-doped semiconductor nanomaterials [35, 36]. Electrochemical responses with calcined $ZnFe_2O_4$ NRs modified electrode substrate are assessed from the different concentrations (0.01 μM to 10.0 M) of formaldehyde, which is exposed in the Figure 4c. It presents the current changes of fabricated films as a function of formaldehyde concentration at room condition. It is also observed that at low to high concentration of target formaldehyde, the current response is enhanced gradually. As confirmed in Figure 4c that with increasing the concentrations of formaldehyde, a gradually change in current was noticed which shows that the conductivity of $ZnFe_2O_4$ NRs modified AgE electrode was considerably increased with increasing the concentration of target chemical. This phenomenon can be attributed to the enhance in ions concentrations and fast communication of electron throughout reaction [37, 38]. The noticeable current changes at higher potential range (potential, +1.0V to +1.5V) based on analyte concentration are magnified and displayed clearly in Figure 4d. A large range of formaldehyde concentration is preferred to inspect the apparent analytical limit, which is calculated in 0.01 μM to 10.0 M. The calibration curve was plotted from the variation of formaldehyde concentrations, which is shown in Figure 4e. The sensitivity is calculated from the calibration curve, which is close to 4.10 ± 0.05 $\mu Acm^{-2}mM^{-1}$. The linear dynamic range of this sensor displays from 0.01 μM to 0.1 M (linearity, R = 0.9406) and the detection limit was considered as 0.0089 μM [3 × noise (N)/slope(S)]. Over the 0.1 M concentration of formaldehyde, the fabricated spinel $ZnFe_2O_4$ NRs sensor becomes saturated. The first phase appears to be the most sensitive section for formaldehyde detection. The fabricated sensors would be constructive at lower concentration region. The saturated area in higher concentration might be owing to the un-accessibility or un-availability of free NR sites for formaldehyde adsorption. The appearance of two phases indicates that the surface reactions are varied at different concentrations. At lower concentration, physic-sorption would take part in key-role, while at higher concentration, chemi-sorption process would be dominant resulting in saturation of the doped semiconductor sensor performances [39, 40].

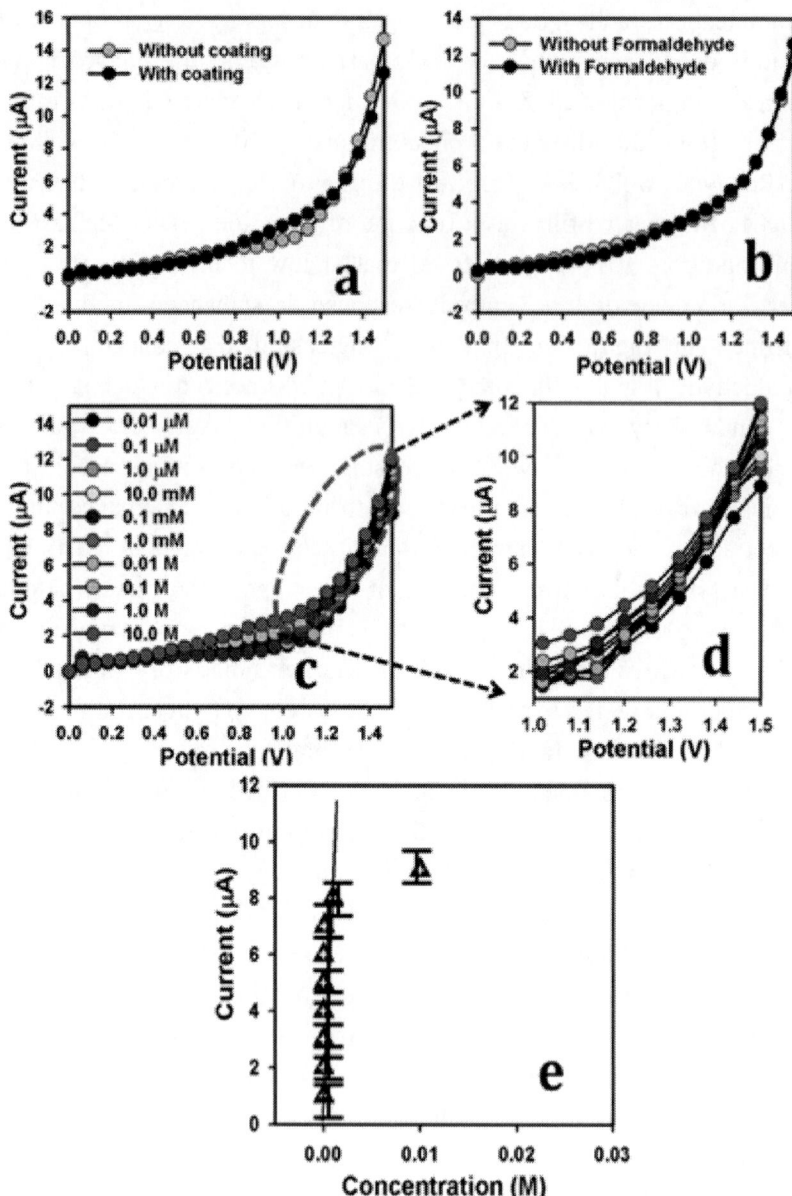

Figure 4. Electrochemical responses of (a) without and with NRs coating of AgE; (b) without and with 0.01 μM formaldehyde in 10.0 mL PBS solution; (C) concentration variations (0.01 μM to 10.0 M) of formaldehyde analyte; (d) magnified view of concentration variation from +1.0 to +1.5V; and (e) calibration plot of spinel $ZnFe_2O_4$ NRs fabricated on AgE surfaces. Potential was taken between 0.0 to +1.5V.

The formaldehyde chemical sensor based on spinel $ZnFe_2O_4$ NRs is displayed good reproducibility and stability for over two weeks and no significance changes in chemi-sensor responses are found. After two weeks, the chemi-sensor response with NRs was slowly decreased, which may be due to the weak-interaction between fabricated NRs active surfaces and formaldehyde chemicals. The significant result was obtained by hydrothermally prepared spinel $ZnFe_2O_4$ NRs, which can be applied as efficient electron mediators for the improvement of proficient chemi-sensors.

Generally, the resistance value of the transition-doped-semiconductor electrodes/sensors are reduced with increasing active surface area at room conditions, due to the basic features of the semiconductor nanomaterials [41]. In fact, oxygen (O_2) adsorption displays a significant liability in the electrical behaviors of the $ZnFe_2O_4$ NRs (n-type semiconductor) spinel structures. Oxygen ion (O_2^-) adsorption eliminates the conduction electrons and increases the resistance of $ZnFe_2O_4$ NRs. Active oxygen species (i.e., O_2^- and O^-) are adsorbed on the calcined material surface at room temperature, and the quantity of such chemisorbed oxygen species strongly depend on the structural properties. At room condition, O_2^- is chemisorbed, while in NRs morphology, both O_2^- and O^- are chemisorbed, and the O_2^- disappears rapidly [42, 43]. Here, formaldehyde sensing mechanism of $ZnFe_2O_4$ NRs sensor is based on the semiconductors oxides, which is held due to the oxidation or reduction of the semiconductor NRs. According to the dissolved O_2 in bulk-solution or surface-air of the adjacent atmosphere, the following reactions (5) & (6) are held in the reaction medium.

$$e^- (ZnFe_2O_4 \text{ NRs}) + O_2 \rightarrow O_2^- \qquad (5)$$

$$e^- (ZnFe_2O_4 \text{ NRs}) + O_2^- \rightarrow 2O^- \qquad (6)$$

These reactions are accomplished in bulk-system or air/liquid interface or neighboring atmosphere due to the small carrier concentration, which enhanced the resistance. The formaldehyde sensitivity towards spinel $ZnFe_2O_4$ NRs could be attributed to the high oxygen lacking conducts to increase oxygen adsorption. Larger the amount of oxygen adsorbed on the NR-sensor surface, larger would be the oxidizing potentiality and faster would be the oxidation of formaldehyde. The reactivity of formaldehyde would have been very large as compared to other chemical with the surface under identical condition [44, 45]. When formaldehyde (HCHO) reacts with the adsorbed oxygen on the sensor surface, it oxidized to carbon dioxide and water, releasing free electrons (2e⁻) into the conduction band, which could be articulated through the following reactions (7).

$$HCHO_{(ads)} + 2O^-_{(ads)} \rightarrow CO_2 + H_2O + 2e^-(C.\ B.) \qquad (7)$$

These reactions related to oxidation of the reducing carriers in presence of semiconductor nanomaterials. These techniques improved the carrier concentration and hence decreased the resistance on exposure to reducing liquids/analytes. At the room condition, the introducing of metal oxide surface to reduce liquid/analytes results in a surface mediated adsorption process. The eradication of iono-sorbed oxygen enhances the electron concentration and therefore the surface conductance of the film [46]. The reducing analyte (HCHO) provides electrons to $ZnFe_2O_4$ NRs surface. Accordingly, the resistance is reduced as well as conductance is increased. This is the cause why the analyte response (current) intensified with increasing potential. Thus the electrons are contributed for rapid increase in conductance of the thick film. The spinel $ZnFe_2O_4$ NRs unusual regions dispersed on the surface would improve the ability of material to absorb more oxygen species giving high resistance in air ambient, which is presented in Figure 5.

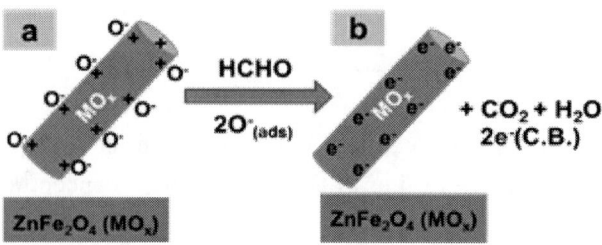

Figure 5. Mechanism of spinel $ZnFe_2O_4$ NR formaldehyde sensors at room conditions.

CONCLUSION

Due to several potential applications of transition metal-doped semiconductor nanomaterials and the excessive benefits of hydrothermal synthesis in cost and environmental impact, a well-crystalline spinel $ZnFe_2O_4$ NR were prepared by simple hydrothermal method at low temperature. The detailed structural, morphological and optical characterizations confirmed that the synthesized nanorods are well-crystalline with spinel face-centered cubic and possessing good optical properties. These NRs exhibited higher sensitivity and lower detection limit with good linearity in short response time and can thus efficiently employed as sensor for the fabrication of formaldehyde chemical sensor. This novel attempt is instigated a well-organized technique of efficient chemical sensor development for environmental hazardous pollutants and biomedical health-care fields in large scale.

REFERENCES

[1] C. Yao, Q. Zeng, G. F. Goya, T. Torres, J. Liu, H. Wu, M. Ge, Y. Zeng, Y. Wang, J. Z. Jiang, $ZnFe_2O_4$ Nanocrystals: Synthesis and Magnetic Properties. *J. Phys. Chem. C.* 111 (2007) 12274-12278.

[2] W. B. InglerJr, J. P. Baltrus, S. U. M. Khan, Photoresponse of p-Type Zinc-Doped Iron(III) Oxide Thin Films. *J. Am. Chem. Soc.* 126 (2004) 10238-10239.

[3] J. Haetge, C. Suchomski, T. Brezesinski, Ordered Mesoporous MFe_2O_4 (M = Co, Cu, Mg, Ni, Zn) Thin Films with Nanocrystalline Walls, Uniform 16 nm Diameter Pores and High Thermal Stability: Template-Directed Synthesis and Characterization of Redox Active Trevorite. *Inorg. Chem.* 49 (2010) 11619-11626.

[4] J. Wang, P. F. Chong, S. C. Ng, L. M. Gan, Microemulsion processing of manganese zinc ferrites. *Mater. Lett.* 30 (1997) 217-221.

[5] M. Grigorova, H. J. Blythe, V. Blaskov, V. Rusanov, V. Petkov, V. Masheva, D. Nihtianova, L. M. Martinez, J. S. Munoz, M. Mikhov, *Magn. Magn. Mater.* 183 (1998) 163-168.

[6] C. H. Yang, H. J. Lee, Y. B. Kim, S. J. Han, Y. H. Jeong, N. O. Birge, Magnetoresistance in Fe and Cu co-doped ZnO thin films. *Physica B.* 383 (2006) 28-30.

[7] F. Grasset, N. Labhsetwar, D. Li, Synthesis and magnetic characterization of zinc ferrite nanoparticles with different environments: powder, colloidal solution, and zinc ferritesilica core-shell nanoparticles. *Langmuir.* 18 (2002) 8209-8216.

[8] H. Deng, X. Li, Q. Peng, X. Wang, J. Chen, Y. Li, Monodisperse magnetic single-crystal ferrite microspheres. *Angewandte Chemie.* (Inter. Ed.). 44 (2005) 2782-2785.

[9] X. Niu, W. Du, W. Du, Preparation and gas sensing properties of ZnM_2O_4 (M = Fe, Co, Cr). *Sens. Actuator B.* 99 (2004) 405-409.

[10] J. A. Toledo-Antonio, N. Nava, M. Martınez, X. Bokhimi, Correlation between the magnetism of non-stoichiometric zinc ferrites and their catalytic activity for oxidative dehydrogenation of 1-butene. *App. Cat. A.* 234 (2002) 137-144.

[11] M. M. Rahman, A. Jamal, S. B. Khan, M. Faisal. Fabrication of Highly Sensitive Ethanol Chemical Sensor Based on Sm-Doped Co_3O_4 Nano-Kernel by Solution Method. *J. Phys. Chem. C.* 115 (2011) 9503-9510.

[12] M. M. Rahman, A. Jamal, S. B. Khan, M. Faisal, Cu-doped ZnO Based Nanostructured Materials for Sensitive Chemical Sensor Applications. *ACS. App. Mat. Inter.* 3 (2011) 1346-1351.

[13] M. Kobayashi, H. Shirai, M. Nunokawa, "Estimation of multiple-cycle desulfurization performance for extremely low concentration sulfur removal with sorbent containing zinc ferrite-silicon dioxide composite powder. *Energy. Fuel.* 16 (2002) 1378-1386.

[14] M. M. Rahman, A. Jamal, S. B. Khan, M. Faisal. Highly sensitive ethanol chemical sensor based on Ni-doped SnO_2 nanostructure materials. *Biosens. Bioelectron.* 28 (2011) 127-134.

[15] J. Fu, D. Gao, Y. Xu, Z. Yan, D. Xue, One-step process to fabricate Fe core/Fe-dimethylsulfoxide shell coaxial nanocables. *Chem. Mat.* 20 (2008) 2016-2019.

[16] G. Zhang, C. Li, F. Cheng, J. Chen, $ZnFe_2O_4$ tubes: synthesis and application to gas sensors with high sensitivity and low-energy consumption. *Sens. Actuator B.* 120 (2007) 403-410.

[17] B. Kazinczy, L. Kótai, I Gács, I. E. Sajó, B. Sreedhar, K. Lázár, Study of the Preparation of Zinc(II) Ferrite and ZnO from Zinc- and Iron-Containing Industrial Wastes. *Ind. Eng. Chem. Res.* 42 (2003) 318-322.

[18] S. R. Lee, M. M. Rahman, M. Ishida, K. Sawada, Fabrication of Highly Sensitive Penicillin Sensor Based on Charge Transfer Techniques. *Biosen. Bioelectron.* 24 (2009) 1877-1881.

[19] S. R. Lee, M. M. Rahman, M. Ishida, K. Sawada, Development of highly sensitive acetylcholine sensor based on acetylcholine by charge transfer techniques esterase using smart biochips. *Trends in Anal. Chem.* 28 (2009) 196-199.

[20] J. D. A. Gomes, M. H. Sousa, F. A. Tourinho, R. Aquino, G. J. D. Silva, J. Depeyrot, E. Dubois, R. Perzynski, Synthesis of Core-Shell Ferrite Nanoparticles for Ferrofluids: Chemical and Magnetic Analysis. *J. Phys. Chem. C.* 112 (2008) 6220-6227.

[21] M. M. Rahman, A. Umar, K. Sawada, Development of Amperometric Glucose Biosensor Based on Glucose Oxidase Enzyme Immobilized with Multi-Walled Carbon Nanotubes at Low Potential. *Sens. Actuator B.* 137 (2009) 327-332.

[22] W. Jiang, Z. Cao, R. Gu, X. Ye, C. Jiang, X. Gong, A simple route to synthesize $ZnFe_2O_4$ hollow spheres and their magnetorheological characteristics. *Smart. Mater. Struct.* 18 (2009) 125013-12016.

[23] T. Sato, K. Sue, W. Suzuki, M. Suzuki, K. Matsui, Y. Hakuta, H. Hayashi, K. Arai, S. Kawasaki, A. Kin, T. Hiaki, Rapid and Continuous Production of Ferrite Nanoparticles by Hydrothermal Synthesis at 673 K and 30 MPa. *Ind. Eng. Chem. Res.* 47 (2008) 1855-1860.

[24] N. Han, L. Chai, Q. Wang, Y. Tian, P. Deng, Y. Chen, Evaluating the doping effect of Fe, Ti and Sn on gas sensing property of ZnO. *Sens. Actuator B.* 147 (2010) 525-530.

[25] Li, Y., Yi, R., Yan, A., Deng, L., Zhou, K., Liu. X. Facile synthesis and properties of $ZnFe_2O_4$ and $ZnFe_2O_4$/polypyrrole core-shell nanoparticles. *Sol. Stat. Sci.* 11 (2009) 1319-1324.

[26] T. C. Damen, S. P. S. Porto, B. Tell, Raman effect in zinc oxide. *Phys. Rev.* 142 (1966) 570-574.

[27] C. Bundesmann, N. Ashkenov, M. Schubert, D. Spemann, T. Butz, E. M. Kaidashev, M. Lorenz, M. Grundmann, Raman scattering in ZnO thin films doped with Fe, Sb, Al, Ga, and Li, *Appl. Phys. Lett.* 83 (2003) 1974-1976.

[28] A. Umar, M. M. Rahman, S. H. Kim, Y. B. Hahn, Zinc oxide nanonail based chemical sensor for hydrazine detection. *Chem. Commun.* (2008) 166-168.

[29] C. Yao, Q. Zeng, G. F. Goya, T. Torres, J. Liu, H. Wu, M. Ge, Y. Zeng, Y. Wang, J. Z. Jiang, $ZnFe_2O_4$ Nanocrystals: Synthesis and Magnetic Properties. *J. Phys. Chem.* C 111 (2007) 12274-12278.

[30] R. Wang, D. Zhang, Y. Zhang, C. Liu, Boron-Doped Carbon Nanotubes Serving as a Novel Chemical Sensor for Formaldehyde. *J. Phys. Chem. B*, 110 (2006) 18267-18271.

[31] S. M. Chen, M. H. Wu, R. Thangamuthu, Preparation, Characterization, and Electrocatalytic Properties of Cobalt Oxide and Cobalt Hexacyanoferrate Hybrid Films. *Electroanalysis* 20 (2008) 178-184.

[32] L. Zhuo, Y. Huang, M. S. Cheng, H. K. Lee, C. S. Toh, Nanoarray Membrane Sensor Based on a Multilayer Design For Sensing of Water Pollutants. *Anal. Chem.* 82 (2010) 4329-4332.

[33] A. Hierlemann, R. Gutierrez-Osuna, Higher-Order Chemical Sensing. *Chem. Rev.*, 108 (2008) 563-613.

[34] L. Feng, C. J. Musto, K. S. Suslick, A Simple and Highly Sensitive Colorimetric Detection Method for Gaseous Formaldehyde. *J. Am. Chem. Soc.*, 132 (2010) 4046-4047.

[35] X. Lai, D. Wang, N. Han, J. Du, J. Li, C. Xing, Y. Chen, X. Li, Ordered Arrays of Bead-Chain-like In_2O_3 Nanorods and Their Enhanced Sensing Performance for Formaldehyde. *Chem. Mater.* 22 (2010) 3033-3042.

[36] M. M. Rahman, A. Jamal, S. B. Khan, M. Faisal. A. M. Asiri, Fabrication of Highly Sensitive Acetone Sensor Based on Sonochemically Prepared As-grown Ag_2O Nanostructures. *Chem. Eng. J.* 192 (2012) 122-128.

[37] L. Mai, L. Xu, Q. Gao, C. Han, B. Hu, Y. Pi, Single β-$AgVO_3$ Nanowire H_2S Sensor. *Nano Lett.,* 10 (2010) 2604-2608.

[38] S. B. Khan, M. Faisal, M. M. Rahman, A. Jamal, Low-temperature Growth of ZnO Nanoparticles: Photocatalyst and Acetone Sensors. *Talanta* 85 (2011) 943-949.

[39] M. Liu, G. Zhao, Y. Tang, Z. Yu, Y. Lei, M. Li, Y. Zhang, D. Li, A Simple, Stable and Picomole Level Lead Sensor Fabricated on DNA-based Carbon Hybridized TiO_2 Nanotube Arrays. *Environ. Sci. Technol.,* 44 (2010) 4241-4246.

[40] S. B. Khan, M. Faisal, M. M. Rahman, A. Jamal, Exploration of CeO_2 nanoparticles as a chemi-sensor and photo-catalyst for environmental applications. *Sci. Tot. Environ.* 409 (2011) 2987-2992.

[41] P. Song, H. W. Qin, L. Zhang, K. An, Z. J. Lin, J. F. Hu, M. H. Jiang, The structure, electrical and ethanol-sensing properties of $La_{1-x}Pb_xFeO_3$ perovskite ceramics with x ≤ 0.3. *Sens. Actuator B: Chem.* 104 (2005) 312-317.

[42] N. Han, L. Chai, Q. Wang, Y. Tian, P. Deng, Y. Chen, Evaluating the doping effect of Fe, Ti and Sn on gas sensing property of ZnO. *Sens. Actuator B.* 147 (2010) 525-530.

[43] T. J. Hsueh, C. L. Hsu, S. J. Chang, I. C. Chen, Laterally grown ZnO nanowire ethanol gas sensors. *Sens. Actuator B.* 126 (2007) 473-477.

[44] B. Tao, J. Zhang, S. Hui, L. Wan, An amperometric ethanol sensor based on a Pd-Ni/SiNWs electrode. *Sens. Actuator B* 142 (2009) 298-302.

[45] E. Wongrat, P. Pimpang, S. Choopun, An amperometric ethanol sensor based on a Pd-Ni/SiNWs electrode. *App. Surf. Sci.* 256 (2009) 968-972.

[46] S. Mujumdar, Synthesis and characterization of SnO_2 films obtained by a wet chemical process. *Mat. Sci. Poland.* 27 (2009) 123-128.

In: A Comprehensive Guide to Formaldehyde   ISBN: 978-1-53619-465-4
Editor: Natasja A. Bach   © 2021 Nova Science Publishers, Inc.

Chapter 4

THERMAL STUDIES ON STYRYL MODIFIED RESORCINOLIC RESIN - FORMALDEHYDE DONOR ADHESION SYSTEMS FOR RUBBER INDUSTRY

J. Dhanalakshmi[1], PhD, S. Siva Kaylasa Sundari[1] and C. T. Vijayakumar[2,], PhD*

[1]Department of Chemistry, Kamaraj College of Engineering and Technology (Autonomous), K.Vellakulam, India
[2]Department of Polymer Technology, Kamaraj College of Engineering and Technology (Autonomous), S.P.G.C. Nagar, K. Vellakulam, India

ABSTRACT

Reinforcing rubber products were made using rubber compounded wih methylene acceptor and methylene donor. Industrial materials like resorcinol formaldehyde resins (RF) and hexamethoxymethylmelamine

---

[*] Corresponding Author's E-mail: ctvijay22@yahoo.com.

(HMMM) are used as methylene acceptor and methylene donor respectively. RF resin possesses excellent usable properties, but they have high hygroscopicity. This propery increases fuming at the time of rubber compounding and handling which is a inherent problem of RF resins to be used in rubbers. Styryl modified resorcinol formaldehyde resin (R2) and HMMM are used in tire industries as an enhanced rubber compounding material. The thermal studies of structurally modified resorcinol formaldehyde resins (alkyl/ aralkyl) are scarce. Styryl group reduces the hygroscopic property of RF resin and resin volatility due to the presence of lower levels of free resorcinol (1-5%) in the R2. The curing characteristics of the pure R2, pure HMMM and R2:HMMM blends (80:20, 60:40, 50:50, 40:60 and 20:80 weight ratios) were investigated using Differential Scanning Calorimetry (DSC) and it was found that the increase in HMMM amount in the blends alter the curing behaviour of pure R2. The amount of heat liberated during the thermal curing of these blends decreased when HMMM content increased in the blends. Thermal properties of thermally cured materials were investigated using thermogravimetric analysis (TGA). TG results indicated that the polymers from the blends were thermally more stable compared to the pure resin R2. Among the materials investigated, the 60:40 and 50:50 weight ratio R2:HMMM blends showed the highest char yield (~36.0%) at 750 °C. The Fourier Transform Infrared Spectroscopic (FTIR) studies of these materials revealed the formation of quinone methide structures during the reaction between the R2 and HMMM and the interaction existing between R2 and HMMM is explicit.

Keywords: aralkyl modified resorcinol-formaldehyde resin, hexamethoxymethylmelamine, DSC, TGA, FTIR

INTRODUCTION

Tires are indispensable product for all the vehicles (Figure 1). By addition of new vehicles, every year there is a huge leap for tire demands. The global tire market is expected to witness significant growth over the forthcoming years. At the same time it has become more complex than ever. At the same time it needs to meet today's stringent requirements in terms of quality, safety and increasing internationalization. A good adhesion between the rubber of a tire and the reinforcing polymeric cord is crucial for the overall tire performance and safety. Polymeric cords are commonly coated with resorcinol formaldehyde resin in order to obtain the desired adhesion.

Figure 1. Tires for different vehicles.

Geometrically a tire is a torus and mechanically a tire is a flexible membrane pressure container (or) a loop of air. Structurally a tire is a high performance composite and chemically a tire consists of materials from long chain macromolecules. The process in tire manufacturing involves (i) mixing process (mixing of raw materials) (ii) manufacturing process (fabrication of various parts of the tire) (iii) tire building (assembling the tire parts) (iv) vulcanization process (tire curing) (v) inspection process (visual and physical inspections of tire) and all tires are then dispatched to the market and are guaranteed as being of a highly reliable quality.

The main compounding ingredients are vulcanizing agents, accelerators, anti-oxidants, anti-ozonants, activators, processing aids and adhesion promoters. Among the compounding ingredients adhesion promoters play a vital role and adhesion between steel cord and rubber [1] is carried out with the help of dry bonding agents.

Dry bonding agents are specially rubber chemicals, comprising an amalgamation between methylene acceptor and donor. The two well-known methylene acceptors are resorcinol and resorcinolic resin (RF) (Figure 2) [2].

Resorcinol is an ideal material for rubber industry. It is a well-known bonding material and already proven as an adhesion promoter for bonding steel cords and synthetic fibres such as polyester, nylon, rayon, aramid, glass and rubber compounds to rubbers. Moreover, resorcinol and resorcinolic resins are considered as high polar molecules in rubber compounds.

Generally, to enhance the properties of the materials, it may be subjected to structural modification of existing materials and/or blending of the materials. So modification is needed in resorcinolic resins for special applications and to reduce the cost associated with a RF resin [3-5]. Resorcinol based compounds and resins are predominantly used as the methylene acceptors in the steel skim rubber compound formulations as adhesion promoters.

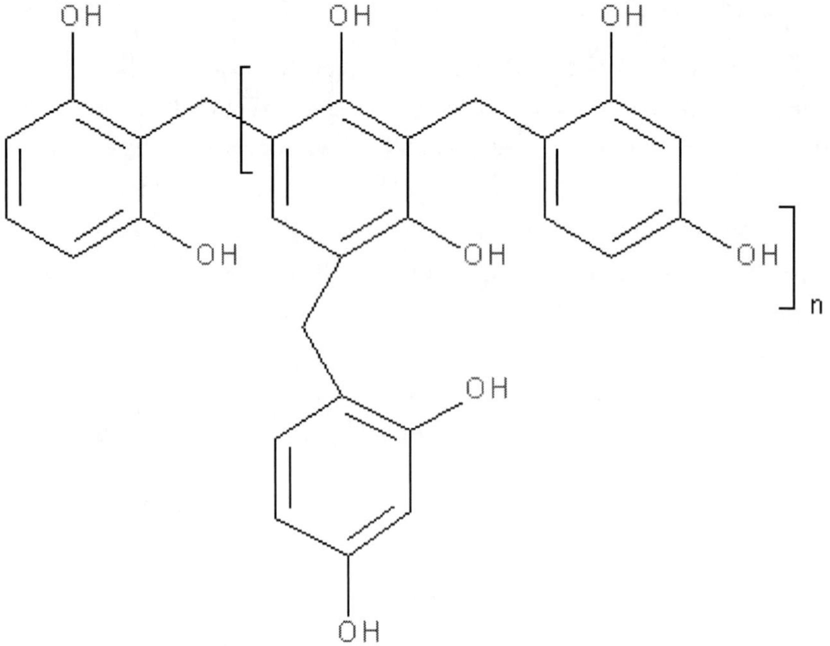

Figure 2. Structure of RF resin.

The problem with RF resins is its hygroscopic nature and become sticky (agglomerated) leading to difficulties in handling RF resins at the tire plants. Hence to reduce the hygroscopic property of resin, to lower free resorcinol monomer in the resin, to reduce resin volatility, to improve the compatibility with rubber, to enhance the mechanical and dynamic properties of cured rubber compounds and to improve the adhesion properties of tire cords with rubbers, structurally modified RF resin is used [6-8].

Table 1. Hygroscopic property of different RF resins

Time	Moisture pickup (%): Hygroscopic property of different RF resins	
	RF resin	Styryl modified RF resin
1 Day	1.1*	0.4
3 Days	2.2*	0.7
One week	2.9*	0.8

* Agglomerated.

Styryl groups in the RF resin (Figure 3) is considered to reduce the hygroscopic property of RF resin (Table 1) and such modified RF resins generally contain low free resorcinol compared to the straight RF resin systems.

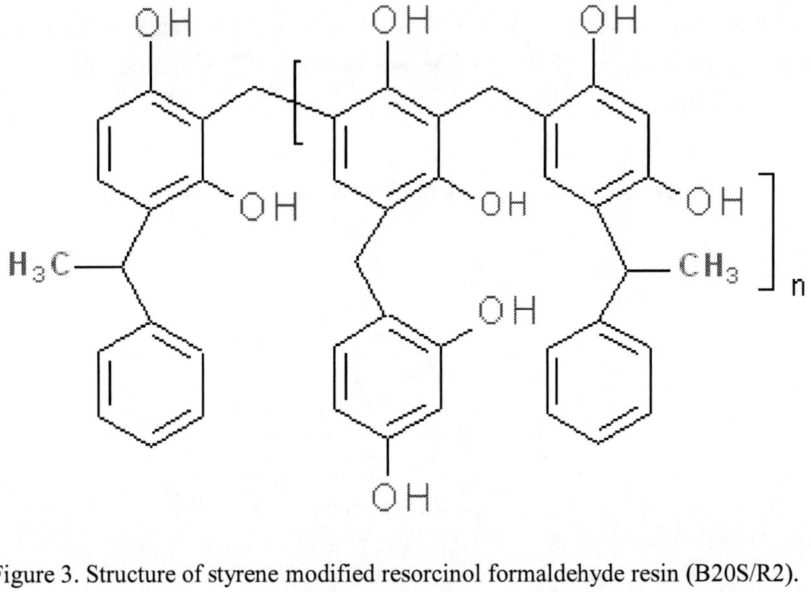

Figure 3. Structure of styrene modified resorcinol formaldehyde resin (B20S/R2).

Hence in the present investigation resorcinol formaldehyde resin is blended in different weight ratios with hexamethoxymethylmelamine. The thermal curing behaviour of these materials were studied using DSC. The thermal stabilities of the thermally cured materials were investigated using TGA and the results were presented and discussed.

EXPERIMENTAL

Materials

Styryl modified resorcinol formaldehyde resin (B20S or R2) and Hexamethoxy-methylmelamine (HMMM) were got as gift samples from Techno Waxchem Pvt. Ltd., Kolkata-700 046, India.

Preparation of Blends

The R2 was separately blended with HMMM (80:20, 60:40, 50:50, 40:60, 20:80 weight ratios) (Table 2) in an agate mortar and the mixture was ground repeatedly to have effective mixing. The mixture was then dried in a vacuum oven and preserved for polymerization.

Table 2. Composition of R2 and HMMM blends

Sample Code	Weight Ratio (R2:HMMM)	Composition	Weight (g)
A2	80:20	4g of R2 + 1g of HMMM	5
B2	60:40	3g of R2 + 2g of HMMM	5
C2	50:50	2.5g of R2 + 2.5g of HMMM	5
D2	40:60	2g of R2 + 3g of HMMM	5
E2	20:80	1g of R2 + 4g of HMMM	5

R2 = Styryl modified resorcinol formaldehyde resin and HMMM = Hexamethoxymethylmelamine.

Thermal Curing of the Materials

The materials pure R2, pure HMMM and R2:HMMM blends were taken in separate micro test tubes and flushed with dry oxygen free nitrogen. The material was thermally polymerized at 150 °C for 3 h. After the polymerization, test tubes were cooled and the samples were removed from the micro test tubes, ground to coarse powder packed and stored for further analysis.

Methods

The differential scanning calorimetric (DSC) curves for materials were recorded in a TA Instruments DSC Q20. The sample (2 - 3 mg) was weighed, placed in the non-hermatic aluminium pan and sealed with aluminium lid. The samples were heated from ambient temperature to 300 °C at a heating rate (β) of 10 °C/min in dry nitrogen (flow rate = 50 mL/min) atmosphere. The obtained DSC curves were analyzed using the universal analysis 2000 software provided by TA instruments.

Thermogravimetric (TG) analyses were performed using TA Instruments TGA Q50. To avoid the secondary reaction of evolved gases in the TG such as thermal cracking, recondensation and repolymerization reactions, the nitrogen flow was maintained at balance and the samples were 50 mL/min and 60 mL/min respectively. The sample (3 - 4 mg) was weighed in a platinum crucible and was heated from ambient temperature to 800 °C at a rate (β) of 10 °C/min.

The FTIR spectra of the materials were recorded in a Shimadzu-8400S infrared spectrophotometer using KBr pellet technique. The absorption bands in IR spectra were used to identify the functional groups in the materials investigated.

RESULTS AND DISCUSSION

Thermal Studies

The differential scanning calorimetric curves recorded at a heating rate of 10 °C/min for the materials R2, HMMM and R2:HMMM blends are shown in Figure 4. The thermal parameters obtained from the DSC curves namely melting temperature (Tm), enthalpy of fusion (ΔHf), onset (Ts), maximum (Tmax) and endset (Te) temperatures of curing and enthalpy of curing (ΔHc) are presented in Table 3.

Table 3. Curing studies of R2:HMMM blends (β = 10 °C/min)

Sample	Ts (°C)	Tmax (°C)			Te (°C)	Te-Ts (°C)	ΔHc (J/g)
A2	134	-	151	166	175	41	9
B2	132	144	164	176	198	66	35
C2	130	145	157	165	195	65	26
D2	126	138	153	173	198	72	23
E2	123	134	159	173	181	58	20

Pure R2 shows a significant endotherm in the temperature region 63 - 105 °C and the endotherm peak at 87 °C, it may be due to the evaporation of the absorbed water molecules in the material. The softening of the material starts around 143 °C [9]. The material shows broad endotherms in the temperature region 143 - 186 °C and the endotherm peak at 166 °C. Since the resin R2 contains <5 % free resorcinol, the sublimation of resorcinol from the resin and probably the melting of the resin may be taking place at this temperature region. The DSC curve for pure 100% HMMM (clear viscous liquid) is also presented in Figure 4 and shows strong multiple endotherms at around 124 and 144 °C. From the DSC curve, it is possible to conclude that the liquid HMMM evaporates and starts to degrade.

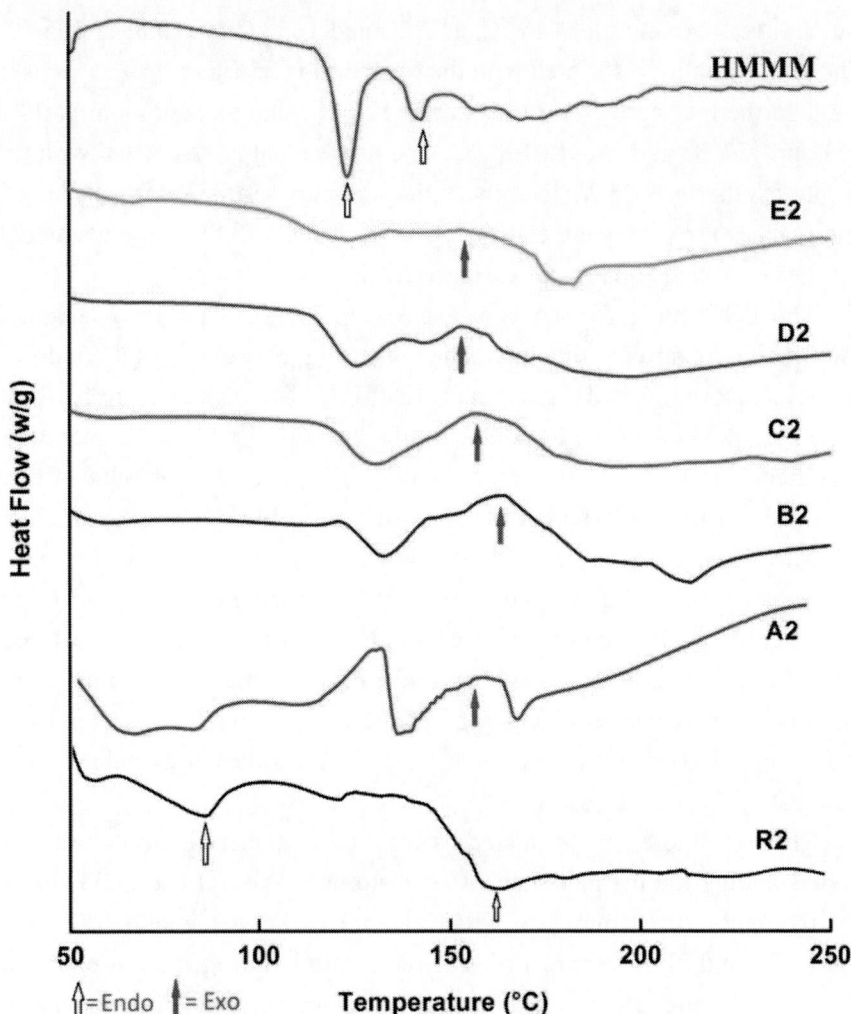

Figure 4. DSC curves of pure R2, pure HMMM and their blends (β = 10 °C/min).

The DSC curves for the R2:HMMM blends are shown in Figure 4. The blend A2 shows a bimodal exotherm. The exotherm starts at 134 °C and ends at 175 °C. The total enthalpy associated with the exotherm (ΔHc) is 9 J/g. The blend B2 shows nearly strong cumulative exotherms. The exotherm starts at 132 °C and shows three maxima at 144 °C, 164 °C and 176 °C and ends at 198 °C. For the blend C2, the exotherm starts at 130 °C and reaches maximum at 145 °C, 157 °C and 165 °C and ends at 195 °C. The total enthalpy associated with the exotherm is 26 J/g.

The thermal curing of D2 starts at 126 °C, attains maximum at 138, 153 and 173 °C and ends at 198 °C. The total enthalpy associated with the curing exotherm is 23 J/g. Similarly thermal curing of E2 starts at 123 °C and reaches three maxima at 134 °C, 159 °C and 173 °C and ends at 181 °C. The observed enthalpy of curing value is 20 J/g.

The DSC curve for the blend shown in Figure 4 provides adequate information regarding the interaction existing between styryl modified resorcinol formaldehyde resin and HMMM. The increase in HMMM content in the blend with R2, decrease the onset of the curing temperature, specifically from A2 to E2 approximately by 10 °C. The amount of heat liberated during the thermal curing of these blends decreases when HMMM content is increased in the blends. Among the materials investigated, B2 required more amount of heat during curing (35 J/g) than the other blends. The compounds R2 and HMMM are also known to affect the cure characteristics of rubber and adhesion may be optimised by varying the percentage of bonding agents. From Table 3, it is found that the increase in HMMM amount in the blends alters the curing behaviour of pure R2.

The TG studies of the materials are one of the major tools used for understanding the thermal stability of materials. The TG and DTG curves for thermally cured pure R2 (PR2) and R2:HMMM blends (PA2, PB2, PC2, PD2 and PE2) were recorded at 10 °C/min in nitrogen atmosphere are shown in Figure 5. The onset degradation temperature (Ts), maximum degradation temperature (Tmax), endset degradation temperature (Te) and char residue obtained at 750 °C for all the samples are tabulated in Table 4. The detailed observations of the thermogravimetric data for all the cured samples are discussed below.

The degradation temperature of thermally cured pure R2 starts at 118 °C, shows multiple stage of maxima at 282, 299 and 320 °C and ends at 443 °C. The char residue for the material PR2 noted at 750 °C is 24%. The TG and DTG curves (Figure 5) clearly indicate that the thermally cured materials PR2, PA2 and PB2 shows multiple mass losses in the temperature region 100 °C to 580 °C with several maxima.

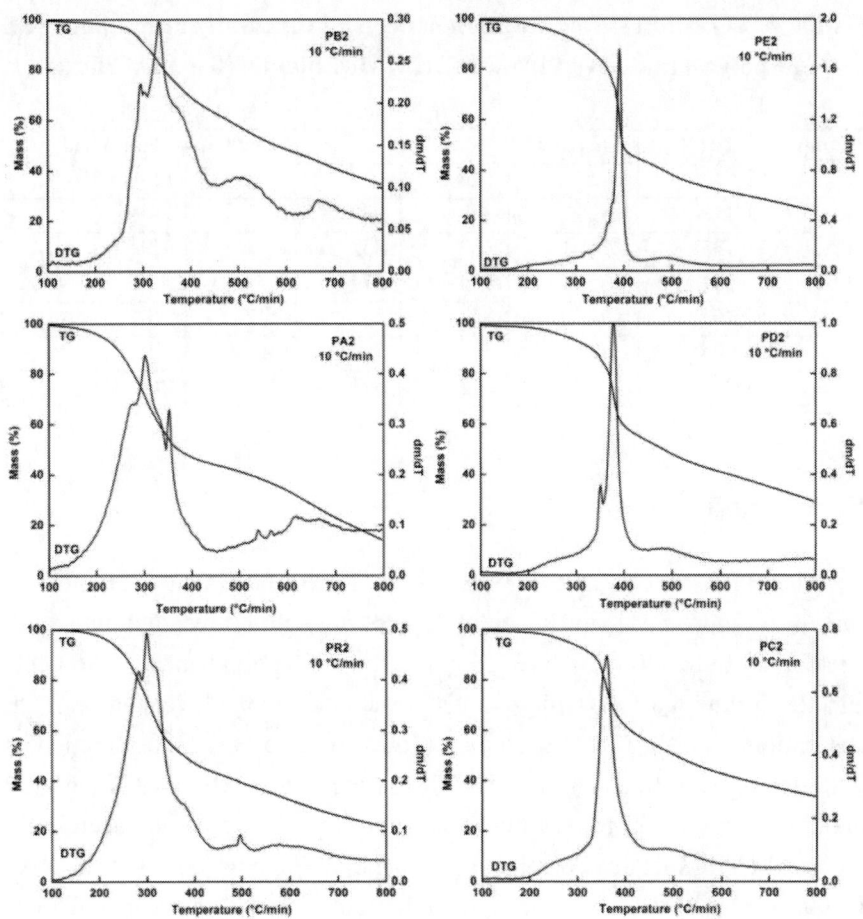

Figure 5. TG and DTG curves for thermally cured pure R2 and R2:HMMM blends.

The polymer from the blends C2, D2 and E2 (PC2, PD2 and PE2) show neat sharper degradation stage followed by a slow degradation process as noted from the DTG curves (Figure 5). From the TG studies of the thermally cured blends, it is possible to understand that as the HMMM content in the blend increases, the degradation pertaining to maximum mass loss is shifted to higher temperatures. It is interesting to note that materials from blends (60:40, 50:50 and 40:60) show the highest char percentage (~36.0%) at 750 °C.

Table 4. TG studies: Degradation aspects of thermally cured pure R2 and polymers derived from R2:HMMM blends (β = 10 °C/min)

Sample	Ts (°C)	Tmax (°C)			Te (°C)	Char Yield (%) at 750 °C
		max1	max2	max3		
PR2	118	282	299	320	443	24
PA2	135	271	302	352	432	18
PB2	201	295	334	374	449	37
PC2	195	-	361	-	447	36
PD2	205	349	378	-	439	32
PE2	208	-	-	391	433	26

FTIR Studies

The FTIR spectra of the resin R2, HMMM and its cured blends are shown in Figure 6. The presence of absorption band corresponding to -OH stretching of phenolic group (3316 cm^{-1}), the absorption bands at 2964 cm^{-1} and 2870 cm^{-1} corresponding to the existence of methylene group, the absorption band for C = C stretching (1605 cm^{-1}, 1508 cm^{-1}), aromatic C = C stretching vibration at 1448 cm^{-1}, C-O stretching at 1019 - 1294 cm^{-1} and the absorption due to trisubstituted benzene ring (839 cm^{-1}) are seen in the recorded FTIR spectra. The absorption band at 700 cm^{-1} corresponds to C-H out-of-plane bending vibration and indicates that there is only one substituent in the benzene ring in the FTIR spectrum of R2 confirmed its structure.

The absorption bands corresponding to C-H stretching of methylene group at 2829, 2933 and 2993 cm^{-1}, N-C-N bending and ring deformation in triazine ring at 1555 cm^{-1}, vibration of methoxy group at 1485 cm^{-1}, antisymmetric deformation of methyl (-CH$_3$) group at 1388 and 1441 cm^{-1}, C-N stretching in triazine ring at 1329 cm^{-1} and 1094 cm^{-1}, C-O stretching in methylol groups at 1022 cm^{-1}, triazine ring at 818 cm^{-1} in the FTIR spectrum of HMMM confirmed its structure.

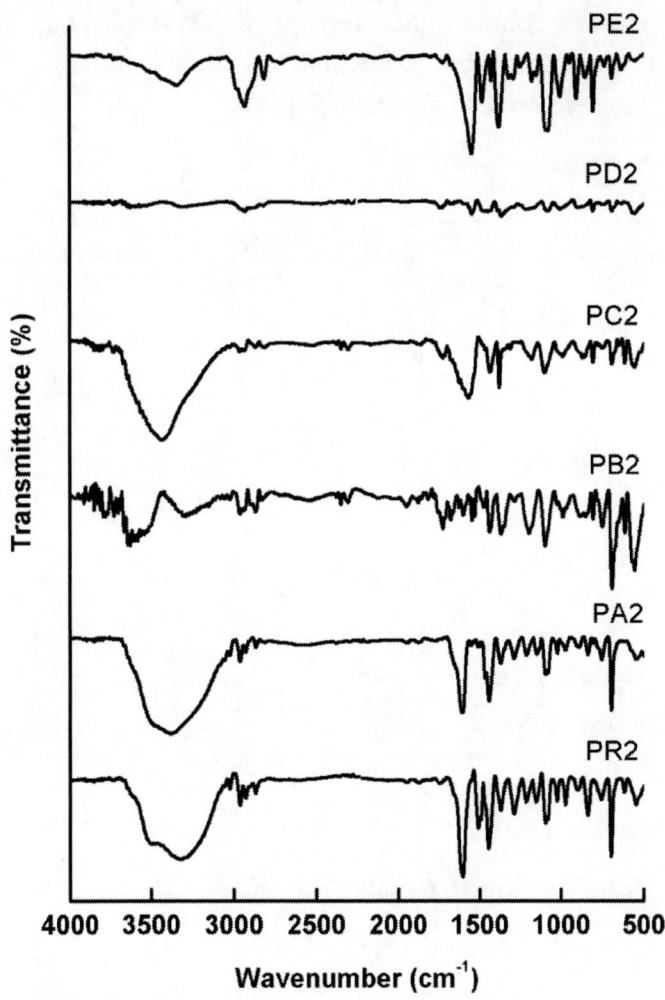

Figure 6. FTIR spectra of thermally cured R2 and R2:HMMM blends.

The thermally cured blends shows decrease in the intensity of the absorption band at 3000-3600 cm^{-1} (phenolic -OH group) indicating the utilization of the phenolic -OH groups during the curing reaction. The weak absorption band at 811 - 1082 cm^{-1} represents the presence of trisubstituted benzene rings in the resin R2. The cured blends show the absorption bands corresponding to N-C-N stretching (1550 cm^{-1}), C-N stretching in triazine ring (1372 - 1471 cm^{-1}) and C-O-C stretching (1155 - 1183 cm^{-1}) which confirms the presence of HMMM unit in the cured material. The multiple weak absorptions occurring in between 1655 and 1849 cm^{-1} are assigned to C = O stretching.

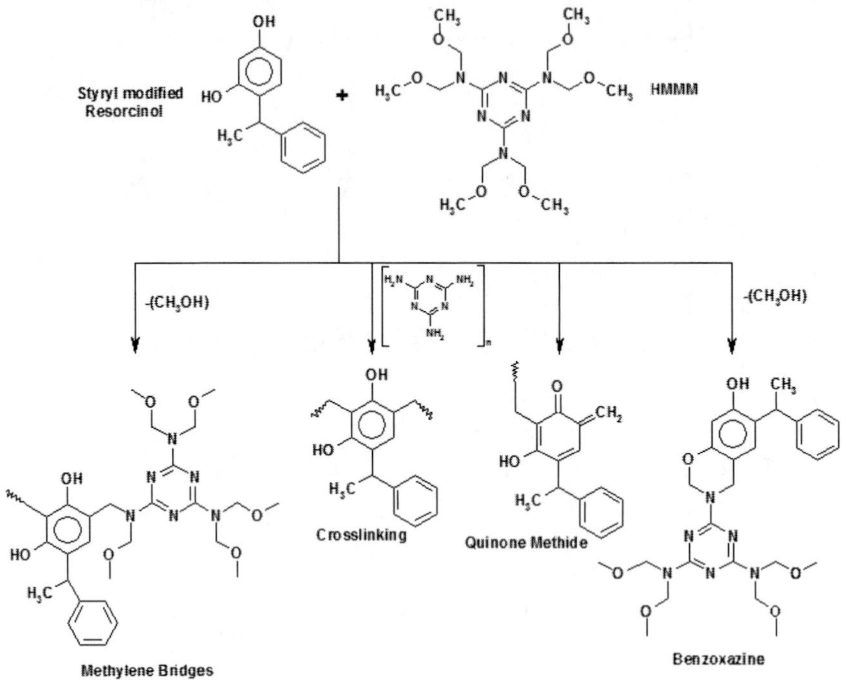

Figure 7. Various possible structures occur from the reaction between R2 and HMMM.

The appearance of new peak at 1678 cm^{-1} confirms the presence of quinone methide structures in the cured material and the intensity of the peak is increased with increase in the HMMM content. In general, the crosslinking reaction between the methylene acceptor and methylene donor leads to the formation of quinone methide and benzoxazine structures [10-11]. The band noted at 989 cm^{-1} is useful to recognize the oxazine ring structure. This band is due to the benzene ring mode of the benzene to which oxazine is attached. When the oxazine ring opens, this mode disappears. It is important to note that disappearance of this mode is not the evidence of oxazine polymerization, but it is a simple indication that the oxazine ring opened. The absence of this band in the thermally cured materials indicates the absence of benzoxazine structures in the blends.

Hence from this preliminary investigation, it is possible to state that quinone methide structures are formed in detectable quantities during the reaction between the methylene acceptor and donor. In the concentration range that has been used and the temperature chosen for the reaction, it is possible to state that benzoxazine units are not formed. In Figure 7 the possible structures resulting from the reaction between methylene acceptor and donor are presented.

In general it can be stated that the mechanism of rubber crosslinking and the chemical structures of the crosslinks are still matters of debate. Ginsburg [12] has suggested a combination of both methylene and chroman crosslinks. With respect to the reaction mechanism, methylene quinones (quinone methides) and benzyl cations have been distinguished as intermediates in ene type or addition reaction respectively. Resorcinolic resins form cross-linked network structures with methylene donors and it will improve rubber compound processability and enhance the adhesion of steel cords to rubbers.

CONCLUSION

In the present investigation styryl modified resorcinolic formaldehyde resin (R2) is taken as the methylene acceptor and HMMM is taken as the methylene donor. The R2 are separately blended with HMMM (80:20, 60:40, 50:50, 40:60, 20:80 weight ratios) and cured thermally at 150 °C for 3 h. The whole investigations were carried out using industrial samples, actually used in the tire industries worldwide. Increase in HMMM amount in the blend with R2, decreases the start of the curing temperature (approximately by 10 °C from A2 to E2). As the HMMM content in the blend increases, the amount of heat liberated during the thermal curing of these blends decreases.

From the TG studies of the thermally cured blends, it is possible to understand that as the HMMM content in the blend increases, the degradation pertaining to maximum mass loss is shifted to higher temperature region. It is interesting to note that the materials PB2, PC2 and PD2 show higher char yields at 750 °C. Although the possibilities of the formations of quinone methide structures and benzoxazine structures exist, the present investigations using FTIR favour the formation of quinone methide structures during the reaction between the methylene acceptor and the methylene donor.

ACKNOWLEDGMENTS

The authors would like to acknowledge the Management and the Principal of Kamaraj College of Engineering and Technology (Autonomous), S.P.G.C. Nagar, K. Vellakulam, India for providing all the facilities to do this work.

REFERENCES

[1] Bhakuni, R. S., and Rye, G. W. (1977). *Bonding tire cord to rubber.* US Patent 4,031,288.

[2] Durairaj Raj, B. (2005). *Resorcinol: chemistry, technology and applications.* Berlin Heidelberg (New York): Springer.

[3] Peter, A. Yurcick. (1968). *Modified resorcinol-formaldehyde adhesive resin and adhesives formed therewith.* US Patent 3,410,818.

[4] Neumann, U. (2001). *Alkylated and/or aralkylated polyhydroxy aromatic compounds and processes for their preparation and use.* US Patent 6, 277, 944 B1.

[5] Harvey, T. Dailey. (2008). *Modified alkylresorcinol resins and applications thereof.* US Patent 2008/0090967 A1.

[6] Hood, R. T., and Lamars, R. M. (1991). *Rubber compounding resin.* US Patent 5,049,641.

[7] Durairaj, B., Peterson, A., Lamers, R. M. and Hood, R. T. (1991). *Aralkyl Modified Resorcinolic Resin.* US Patent 5,021,522.

[8] Yurkevichyute, A. S., Grigor eva L. S. and Vasil ev V. V. (2016). Synthesis of solid resorcinol-formaldehyde resin modified with styrene with the use of a shale phenol fraction with a boiling temperature higher than 270 °C. *Solid Fuel Chemistry,* 50, 64-68.

[9] Durairaj, B., and Lawrence, M. A. (2006). *Modified resorcinol resins and applications thereof.* US Patent 7,074,861 B2.

[10] Knowroski, R. A., Shocker, J. P., Gregg, E. G. and Savoca J. L. (1986). Resorcinol formaldehyde latex (RFL) adhesives and applications. *Rubber Chemistry and Technology.* 59, 328.

[11] Van Gils G. E. (1968). Reaction of resorcinol and formaldehyde in latex adhesives for tire cords. *Industrial and Engineering Chemistry Product Research Division*, 7, 151.

[12] Ginsburg, L. B., Scherschnew, W. A. and Dogadkin, B. A. (1973). *Ber. Akad. Wiss. UdSSR* 152, 335.

INDEX

A

absorption pea, 14, 65, 93
absorption spectrum, 93
accessibility, 48, 59, 69, 72, 99
acetone, 89, 91
acetylcholine, 105
activated carbon, 57
activation energy, 9, 13, 19, 20, 21, 23, 27, 28, 29, 30, 40, 41, 42, 43, 44, 92
active surface area, 101
adhesion, 49, 110, 111, 112, 113, 118, 123
adhesion promoters, 111, 112
adhesion properties, 113
adhesives, 3, 49, 124, 125
adsorbed oxygen, 69, 102
adsorption, 54, 55, 56, 57, 66, 68, 69, 72, 81, 99, 101, 102
advanced isoconversional kinetic methods, 34
Advanced Vyazovkin Method (A-VYZ), 19, 20, 23, 27, 28, 30, 43
aggregation, 51, 53, 68, 69
air quality, 58, 82
aldehydes, 69, 71
aliphatic amines, 38
amine, viii, 2, 7, 14, 33, 37, 57, 81
ammonium hydroxide, 90
aniline, 7, 11, 14, 26, 37, 38
antibody, 53
antitumor, 5
apparent activation energy, 9, 13, 19, 20, 21, 23, 27, 28, 29, 30, 40, 41
aralkyl modified resorcinol-formaldehyde resin, 110
atmosphere, 23, 50, 53, 101, 115, 118
atoms, 5, 54, 66, 72

B

band-gap energy, 94
behaviors, 88, 101
bending, 94, 120
benzene, 5, 14, 15, 25, 120, 122
benzoxazine units, 123
binding energies, 66
biological samples, 73
biomaterials, 72
biopolymer, 84
biopolymers, 3

bisbenzoxazine, vii, 2, 8, 9, 10, 11, 12, 14, 15, 17, 21, 25, 26, 33, 34
bisphenol, vii, 2, 7, 17, 21, 26, 37, 38, 39
blends, ix, 36, 110, 114, 115, 116, 117, 118, 119, 120, 121, 123, 124
bond dissociation energy, 32
bonding, 15, 36, 48, 54, 69, 72, 83, 111, 112, 118
bonding agents, 118
butyl carbitol acetate, 90, 92

C

calibration, ix, 88, 92, 99, 100
calibration plot, ix, 88, 92, 100
carbon dioxide, 51, 97, 102
carbonyl groups, 56
carboxylic acid, 37
carboxylic acids, 37
carcinogenesis, 51
carcinogenicity, 76
catalytic activity, 104
catalytic effect, 21
catalytic properties, 69
cell line, 77
cellulose, 83
chain mobility, 21
challenges, 59, 68
chemical, vii, viii, ix, 3, 7, 21, 37, 44, 47, 48, 54, 57, 58, 64, 69, 88, 89, 91, 92, 94, 97, 99, 101, 102, 103, 104, 106, 108, 123
chemical bonds, 94
chemical industry, 58
chemical inertness, 3
chemical interaction, 89
chemical properties, ix, 70, 88
chemical reactions, 21, 91
chemical sensing, ix, 88, 89, 92, 106
chemical sensor, vii, ix, 88, 89, 97, 101, 103, 104, 106
chemical stability, 37

chemical structures, 123
chemicals, 7, 48, 49, 50, 55, 89, 90, 97, 101, 111
chemisorption, 69
chloroform, 11, 89
cleavage, 32, 33, 34
compounds, vii, 2, 14, 17, 30, 31, 32, 33, 34, 36, 38, 48, 112, 118
condensation, 5, 7, 14
conductance, 71, 102
conduction, 98, 101, 102
conductivity, 70, 99
confinement, 83, 84
consumption, 14, 51, 81
contaminant, 56, 70
contaminated food, 51
contamination, 51, 59
Corrected Flynn – Wall – Ozawa (C-FWO) Method, 42
Corrected Kissinger – Akahira – Sunose (C-KAS) Method, 43
correlation, 50, 53, 66, 69, 73
crystalline, 70, 94, 95, 96, 103
curing process, viii, 2, 3, 13, 21
curing reactions, 17
current response, 89, 97, 98

D

decomposition, 9, 23, 31, 37
degradation, viii, 2, 9, 23, 24, 25, 26, 27, 28, 29, 30, 31, 32, 33, 34, 37, 38, 43, 118, 119, 124
degradation mechanism, 9, 31, 32
degradation process, viii, 2, 26, 31, 119
degree of crystallinity, 96
derivatives, 5, 33, 35, 39, 49
detectable, 60, 68, 70, 123
detection, ix, 48, 51, 58, 59, 61, 66, 67, 68, 70, 72, 73, 84, 88, 89, 97, 98, 99, 103, 106

Index

detection limit, ix, 88, 99, 103
detection system, 58
differential scanning, 21, 36, 37, 40, 115, 116
differential scanning calorimetry, 21, 36, 37, 40
diffusion, 21, 60, 70, 82
disease progression, viii, 47, 53
doping, 106, 107
drug reactions, 74
drugs, 74
dry bonding agents, 111
DSC studies, 17, 18

E

Ea-C, viii, 2, 9, 13, 19, 20, 21, 22, 23
Ea-D, viii, 2, 9, 13, 27, 28, 29, 32
economic growth, 48
electrical properties, 7
electro-catalytic oxidation, 99
electrochemical response, 99, 100
electrode surface, 89, 97, 98
electrodes, ix, 84, 88, 92, 101
electrometer, 90, 92, 97, 98
electron, 21, 66, 68, 69, 72, 98, 101, 102
electron communication, 99
electrons, 69, 98, 101, 102
emission, 59, 67, 68, 71
energy, 13, 20, 21, 22, 28, 29, 32, 41, 42, 44, 67, 84, 94, 95, 105
energy consumption, 105
engineering, 3, 59, 60, 66
environmental change, 54
environmental conditions, 73
environmental impact, 103
environments, viii, 3, 47, 48, 49, 51, 53, 55, 58, 104
ethanol, 10, 11, 89, 91, 104, 107, 108
ethyl acetate, 90, 92
evaporation, 12, 50, 116

evidence, 14, 51, 76, 123
evolution, 33, 34, 68
exchange rate, 85
excitation, 67, 68, 71
exposure, 51, 52, 53, 58, 66, 75, 76, 77, 78, 79, 81, 102

F

fabrication, 70, 73, 84, 103, 111
films, 89, 99, 108
fluorescence, 53, 67, 68, 69, 71
Flynn-Wall-Ozawa Method (FWO), 9, 19, 20, 21, 22, 28, 40, 42, 43
Flynn – Wall – Ozawa Method (FWO), 42
formaldehyde, v, vi, vii, viii, ix, 3, 5, 6, 11, 35, 47, 48, 49, 50, 51, 52, 53, 54, 55, 56, 57, 58, 59, 61, 62, 63, 64, 65, 66, 67, 68, 69, 71, 72, 73, 74, 75, 76, 77, 78, 79, 80, 81, 82, 83, 84, 85, 87, 88, 89, 90, 92, 97, 98, 100, 101, 102, 103, 106, 107, 109, 110, 111, 113, 114, 118, 123, 124, 125
formation, x, 6, 14, 17, 25, 34, 67, 68, 70, 74, 94, 96, 110, 122, 124
Franklinite, 96
free energy, 64
free radicals, 76
Friedman Method (FRD), 19, 20, 28, 42
FTIR studies, 14, 31, 34, 39, 120
FTIR technique, 35

G

gas sensors, 70, 105, 107
geometry, 96
glass transition, 7
global scale, 48
gold nanoparticles, 65
growth, vii, viii, ix, 47, 58, 84, 88, 91, 92, 110
growth mechanism, 91

H

harmful effects, 52, 55, 73
hazardous pollutants, 103
health, vii, viii, 48, 51, 54, 58, 73, 81, 103
health effects, 51, 81
health risks, 73
heat release, 17, 18
heating rate, 12, 17, 18, 19, 22, 23, 24, 25, 30, 41, 42, 44, 115, 116
hexamethoxy- methylmelamine, 110
human, 53, 55, 77, 78
human body, 55
human health, 53
humidity, 10, 50, 52, 57
humoral immunity, 75
hydrazine, 68, 89, 106
hydrogen, 10, 15, 36, 69, 83
hydrothermal, vii, ix, 81, 88, 89, 90, 103, 105
hydrothermal synthesis, 103

I

immunosuppression, 75
industries, vii, ix, 48, 110, 123
infrared spectrophotometer, viii, 2, 115
interface, 66, 72, 101
internationalization, 110
IR spectra, 90, 94, 115
iteration, 22, 42
I-V technique, 97

K

kinetic equations, 13
kinetic methods, viii, 2, 19, 27, 34
kinetic model, 29
kinetic parameters, 9

kinetics, 6, 8, 20, 21, 22, 26, 29, 34, 36, 37, 38, 39, 40, 41, 54
Kissinger-Akahira-Sunose Method (KAS), 9, 19, 20, 21, 22, 28, 41, 43, 44
Kissinger – Akahira – Sunose Method (KAS), 43

L

large dynamic range, ix, 88
lifetime, 30, 31, 44, 45
linear polymers, 6
living environment, 58
low temperatures, 29
luminescence, 69

M

magnetic characteristics, 88
magnetic materials, 88
magnetic properties, 89
manufacturing, vii, 3, 49, 50, 54, 77, 111
material surface, 49, 101
materials, vii, ix, 3, 8, 9, 12, 17, 19, 23, 25, 27, 30, 34, 45, 48, 49, 54, 55, 56, 57, 58, 60, 64, 70, 73, 80, 81, 83, 88, 89, 94, 95, 104, 109, 111, 112, 114, 115, 116, 118, 119, 120, 123, 124
mathematical methods, 9
measurement, 59, 90, 98
mechanical properties, 3, 7, 36, 88
melting temperature, 116
meso-porous structure, 94
metabolic change, 51
metal oxides, 57, 89
methodology, 98
methylene acceptors, 111, 112
methylene donor, vii, ix, 109, 122, 123, 124
micrometer, 69
micronucleus, 77
microspheres, 104

microstructure, 84
molecular dynamics, 83
molecular orientation, 69
molecular structure, 2, 26
molecular vibration, 94
molecular weight, 3, 5, 80
molecules, viii, 2, 25, 51, 57, 59, 64, 66, 67, 68, 69, 71, 72, 112, 116
monomer molecules, 3
monomers, viii, 2, 6, 7, 9, 12, 13, 14, 15, 19, 21, 34, 54
monosaccharide, 67
morphology, 64, 65, 70, 89, 92, 95, 101

N

nanomaterials, 48, 66, 73, 88, 98, 101, 102, 103
nanometer, 69, 72, 95
nanoparticles, 66, 67, 69, 71, 72, 84, 104, 106, 107
nanorod, v, ix, 87, 88, 89, 93
nanostructures, 89, 95
nitrogen, 5, 15, 23, 26, 29, 58, 81, 115, 118
NR-sensor surface, 102
nucleation, 64, 91, 92
nuclei, 53, 92
nucleus, viii, 2

O

oligomers, 6, 8, 14
optical properties, 92, 93, 103
optimization, 54
organic polymers, 57
organs, 51, 53
oxidation, 5, 57, 81, 99, 101, 102
oxidative stress, 51, 76
oxide nanoparticles, 69
oxygen, 5, 29, 69, 72, 94, 101, 102, 115

P

permission, iv, 49, 57, 61, 67, 71
phenol, viii, 2, 3, 6, 7, 21, 33, 34, 125
phenolic compounds, 14, 34
phenolic resins, 3, 7, 37
photocatalytic action, 93
physical properties, 7, 88
pollution, 55, 74
polyamines, 36
polybenzoxazines, 6, 7, 31, 36, 38
polymer, viii, 2, 3, 7, 9, 12, 23, 26, 27, 29, 35, 37, 38, 49, 50, 54, 67, 68, 72, 83, 119
polymer films, 67
polymer materials, 50
polymer properties, 7
polymer systems, 23
polymeric chains, 28
polymeric materials, 2, 9
polymerization, 3, 5, 6, 7, 14, 15, 17, 20, 21, 22, 25, 26, 27, 35, 37, 39, 54, 74, 80, 114, 115, 123
polymerization mechanism, 25
polymerization process, 54
polymerization temperature, 26
polymers, viii, x, 2, 3, 7, 9, 13, 14, 15, 16, 27, 35, 36, 38, 40, 110, 120
preparation, iv, 36, 70, 88, 124

Q

quantification, 60, 61, 89
quinone, x, 110, 122, 123, 124
quinone methide, x, 110, 122, 123, 124

R

Raman spectroscopy, ix, 88, 93, 94
reaction mechanism, 43, 91, 97, 98, 123
reaction medium, 101

reaction rate, 13
reactions, 7, 12, 17, 40, 79, 101, 102, 115
reactivity, 25, 72, 102
recognition, 64, 68, 72, 89, 94
recommendations, iv, 58
requirements, 8, 58, 110
researchers, 56, 58, 66
resins, ix, 3, 5, 7, 37, 39, 48, 49, 58, 77, 109, 112, 113, 123, 124, 125
resistance, 3, 7, 29, 66, 69, 71, 88, 101, 102
resorcinol, ix, 109, 110, 111, 112, 113, 114, 116, 118, 124, 125
resorcinol formaldehyde resin, ix, 109, 111, 113, 114
response, ix, 50, 54, 65, 70, 84, 88, 89, 97, 98, 101, 102, 103
response time, ix, 70, 88, 89, 103
ring-opening addition reaction, 7
rings, 14, 15, 25, 122
room temperature, 10, 11, 50, 91, 101
rubber, vii, ix, 109, 110, 111, 112, 113, 118, 123, 124
rubber compounds, 112, 113
rubber products, vii, ix, 109

S

safety, 58, 110
selectivity, 66, 67, 72, 84
semiconductor, 57, 69, 88, 89, 94, 98, 99, 101, 102, 103
semiconductor nanomaterials, 101, 102, 103
seminiferous tubules, 78
sensing, ix, 48, 64, 65, 66, 69, 72, 84, 88, 89, 92, 98, 101, 104, 106, 107
sensing technologies, 48, 65, 69
sensitivity, ix, 59, 66, 67, 69, 70, 76, 84, 88, 92, 98, 101, 103, 105
sensor, vii, ix, 66, 69, 84, 88, 89, 97, 99, 101, 102, 103, 104, 105, 106, 107, 108
silver, ix, 88, 89, 97

silver electrode, ix, 88, 89, 92, 97
solution, 10, 11, 17, 44, 56, 90, 91, 92, 98, 100, 101, 104
sorption process, 99
species, 48, 76, 101, 102
spinel face-centered cubic, 103
spinel $ZnFe_2O_4$ NRs, 89, 93, 94, 95, 96, 97, 98, 100, 101, 102
stability, ix, 3, 22, 27, 50, 54, 88, 97, 101
stretching, 14, 94, 120, 122
structural defects, 93
structurally modified RF resin, 113
structure, viii, 2, 7, 8, 14, 15, 25, 26, 29, 31, 59, 64, 81, 89, 94, 95, 96, 107, 120, 121, 122
styrene, 113, 125
styryl modified resorcinol formaldehyde resin, ix, 110, 114, 118
surface area, 58, 69, 72, 92, 95, 97, 101
surface properties, 65, 72
surface reaction, 99
surface reactions, 99
symptoms, 55, 80
syndrome, 55, 80
synthesis, vii, 5, 27, 36, 37, 39, 91, 105, 106

T

temperature, viii, 2, 3, 9, 11, 12, 13, 17, 21, 22, 23, 25, 26, 27, 29, 30, 31, 33, 34, 36, 38, 41, 42, 43, 44, 45, 50, 52, 69, 89, 90, 91, 97, 103, 107, 115, 116, 118, 123, 124, 125
TG-FTIR, viii, 2, 12, 31, 32, 33, 35, 39
thermal analysis, 37
thermal decomposition, 38
thermal degradation, viii, 2, 8, 9, 13, 25, 26, 27, 28, 29, 31, 32, 33, 35, 36, 38, 39
thermal lifetime prediction, 44
thermal oxidative degradation, 35
thermal properties, 9, 36, 54

thermal stability, viii, 2, 9, 15, 23, 26, 27, 34, 39, 118
thermograms, 23
thermogravimetric analysis (TGA), viii, x, 2, 12, 23, 34, 35, 39, 110, 114, 115
thermogravimetry, 36
thermosets, viii, 2, 3, 9
thin films, 84, 89, 104, 106
toxicity, 55, 72, 75, 78, 97
trajectory, vii, viii, 47, 58
transition-doped semiconductor, 94, 99

V

vapor, 67, 81, 84
variations, 52, 100

Vyazovkin Method (VYZ), 9, 19, 20, 27, 28, 41, 43

W

water, 7, 11, 58, 66, 73, 81, 90, 92, 94, 97, 102, 116
wave number, 95
weak-interaction, 101
weight ratio, ix, 110, 114, 123

Z

zinc, 90, 103, 104, 106